Dr Richard Elwes is a writer, teacher and researcher in Mathematics and a Visiting Fellow at the University of Leeds. He contributes to *New Scientist* and *Plus Magazine* and publishes research on model theory. Dr Elwes is a committed popularizer of mathematics, which he regularly promotes at public lectures and on radio.

CHAOTIC FISHPONDS AND AND MIRROR UNIVERSES

Richard Elwes

Quercus

CONTENTS

INTRODUCTION

Of all the subjects studied, debated and fought over in the course of human history, I happen to believe that the most fascinating is mathematics. That's a bold claim – perhaps mystifying to readers who were bored or baffled by the subject at school. Well, of course fascination is in the eye of the beholder, and certainly there will be those who need some persuading. I hope this book will go some way towards doing that.

What is irrefutable, however, is that in modern life mathematics is both *important* and *ever-present*. Even the most entrenched maths-hater has an awareness that it plays a central role in today's world, touching our lives in more ways than ever before. But that is where the details are liable to become hazy . . . yes, important, but where *exactly* is it used, and in what ways?

In response, I present in the pages that follow a selection of 35 diverse applications of mathematics. I attempt to unravel some of the principles that underlie aspects of our daily lives, as well as those that inform today's boldest thinkers. We will touch on numerous branches of science, seeing how chaos theory can illuminate our understanding of population growth and why the universe's black holes are – contrary to all expectations – rich in information.

Money – that universal experience of numbers – is explored in the context of options and futures, along with the gambler's intimate (if rocky) relationship with probability. The computing revolution, of course, suffuses the discussion, as does its offspring the Information Age. Most people have a sense that mathematics is deeply involved in these developments – the air is thick with talk of binary code and that now popular word, the 'algorithm'. From delving into the innards of search engines, to investigating the structures of social networks and illuminating the ideas behind programming languages and CGI imagery, the aim is to bring this mathematics out of the shadows.

I want to show how mathematics can be used to solve some profoundly pragmatic problems, helping businesses to run efficiently, highlighting flaws in our democracies, providing valuable ammunition in the fight against disease, and analysing the highs and lows of our economic systems. It can even shine a light on the foibles of human psychology, demonstrating where we are prone to make decisions against our best interests. We will discover ways to tell fact from fiction too, by way of those much-used and much-abused numbers: statistics.

All this extremely *useful* mathematics does much to free the modern mathematician from the ivory tower he, or she, is meant to inhabit. This is not to say that today's mathematicians have lost their imagination or no longer reach for the stars. Indeed, in 1969 Apollo 11's touchdown on the moon was made possible by the analysis of one particularly fiendish mathematical problem. In fact, the stars and planets have provided us with numerous mathematical insights over the centuries, which can now be turned upon the still bigger questions – about the nature of matter and the universe.

I hope that, by the end of this book, readers will have a more precise sense of where mathematics fits into modern life – and, *en route*, that some doubters become devotees of the subject that I find so endlessly, gloriously, fascinating.

KNOWING ME, KNOWING YOU

The mathematical hierarchies of knowledge

Imagine a scene. Sarah and Sam used to be a couple. Today they are meeting again, for the first time since their relationship broke down. Since then, Sarah has found a new boyfriend, but has decided not to tell Sam. Unbeknownst to her, however, Sam already does know, because a mutual friend, Saul, spotted the two together and told Sam.

It sounds as though we have left mathematics for the world of romantic comedy or soap-opera. In fact, the dramas of human relationships provide examples where mathematicians' concepts of different 'orders' of knowledge come into play. These ideas are central to the mathematical genre of game theory and in philosophy. Their practical implications continue to be debated, especially in economics. In a competitive market, possessing the same (or superior) knowledge to one's rivals can make all the difference.

But let's return to the drama. In the language of orders of knowledge, the existence of Sarah's new partner is first-order knowledge: both parties know about it. However, there is an imbalance at a higher level; because Sarah is unaware that Sam knows about her new amour, Sam has got the upper hand – he has a piece of second-order knowledge that Sarah does not. To put it another way, he knows that she knows, but she doesn't know that he knows.

We can now ratchet up the drama and imagine a more complex version of the scene. Sarah has in fact found out that Sam knows about her new boyfriend, because Saul later admits to telling him. Sam is now unaware that Sarah knows he knows (and that Saul broke his promise not to tell Sarah). It may be that Sam tries to entrap her with probing questions, to see if she lies or admits the truth under pressure. But in this version – unlike our original scenario – she is unlikely to lie, knowing as she does that that she would get caught. (In the original she might have done.) Now Sam is at a disadvantage, because Sarah has a piece of third-order knowledge that he does not possess.

Could these orders of knowledge increase, *ad infinitum*? In principle yes, but even in the most convoluted spy story in which double agents double cross other double agents, it is rare to find orders of knowledge beyond four. Higher orders of knowledge become increasingly hard to imagine, but in theory knowledge *can* exist at any order.

Among a group of people, if each of them has knowledge of the same thing – call that x – then x is first-order knowledge. What is more, if the group members know that they all share this first-order knowledge, then they are party to an additional level of knowledge, making x second-order knowledge. In general, when the whole group knows that x is nth-order knowledge, then it becomes $(n + 1)$th-order knowledge. For higher orders of knowledge to come into play, more people need to be involved.

Messy diners and the Emperor's clothes

A number of classic puzzles illustrate orders of knowledge, of which the tale of the messy diners is one. Imagine a grand banquet is taking place in a palace, but when the staff laid the table they forgot to put out the napkins, which are all still in the laundry. So, after the meal, an embarrassing situation arises in which some of the diners are left with food on their faces. No diner can see his or her own face, which presents them with a dilemma. According to the local etiquette, it would be unthinkably impolite for one diner to tell another that they have a dirty face. On the other hand wiping one's face on the tablecloth would also be poor manners. It *could* be worth the risk if the diner could be certain that they had a dirty face, but not otherwise.

After dinner, the butler realizes the situation and finds an imaginative solution. 'Ladies and gentlemen,' he announces 'I am sorry to have to inform you that at least one person around this table has a dirty face. I shall now ring the gong repeatedly. After each ring, any guest who is certain they have a dirty face should wipe it.' By way of this announcement, the butler has elevated a piece of information into the realms of common knowledge. Not only does each member of the group hear it, but each knows that everyone else has heard it too.

The finale to the messy meal depends on the number of people who have food on their faces. If there is just one, since she cannot see anyone else with a dirty face, she must logically immediately deduce from the butler's announcement that the single dirty face is her own. If there are two messy faces around the table, each would see the other's dirty face and be unable to make a deduction about themselves. At the level of first-order knowledge, the butler does not tell them anything they did not already know. However, each of the two messy diners would also make the critical observation that the other does *not* wipe their face after the first gong-beat. From this extra piece of data, each would logically deduce the existence of a second messy eater, who – since they can see everyone else's clean faces – can only be themselves. They will each clean up after the second gong. In fact, there is a rule: if there are n messy eaters, they will all wipe their faces after the nth beat of the gong.

The effect of the butler's announcement has a better-known parallel from children's literature, in the Hans Christian Andersen story *The Emperor's New Clothes*, in which a vain emperor is promised a beautiful set of clothes by two swindlers posing as tailors. They assert that the fabric they use is invisible to anyone stupid or incompetent at their jobs. Not wanting to be thought a fool, the Emperor admires his new 'clothes', while his advisers also coo over their beauty. And as the Emperor parades through the town, supposedly newly attired, he is greeted with reverent admiration. It is only when one child shouts out the obvious – that that the Emperor is naked – that uproar begins.

Andersen's clever story works on many levels. It is a parable about the human propensity to avoid pricking a bubble, however fantastical, in the face of peer pressure and authority. It also depends on the fact that, while every character can see that the Emperor has nothing on, no individual can be completely certain that others are seeing the same thing. The child's announcement elevates that reality to common knowledge – in our terms, knowledge of every order, just as the butler's announcement at the dinner party did – shattering the mass delusion.

The naked emperor also serves as a satire of the absurdities of social convention. It was during his analyses in the late 1960s of how social conventions emerge that the philosopher David Lewis became one of the first to investigate the idea of common knowledge. He suggested that conventions arise as solutions to group-coordination problems. A simple example might be deciding on which side of the road to drive. There is no particular reason to favour either the left or the right, but there certainly is a good reason to want everyone to adopt the same custom on a country's roads. Although the choice behind the convention may be essentially arbitrary, once established it will self-perpetuate. With everyone else driving on the left, no individual will have any incentive to drive on the right.

For a convention to take root requires that it become common knowledge among the population. Subtler examples include language and money. It is very convenient to have tokens that can be exchanged for a variety of goods, but the system will only work if everyone believes that the tokens they earn can be spent on whatever they wish to buy. Every individual citizen requires every other individual citizen to consider the tokens as valid currency.

Is there life on Mars?

One of the first people to analyse the phenomenon from a more mathematical perspective was Robert Aumann, in 1976. His Agreement Theorem unveiled something astonishing: that two 'rational agents' – the conventional description of idealized people assumed to think perfectly logically – cannot agree to disagree about the likelihood of a particular speculative event,

X, such as, say, the existence of life on Mars or the election of a female US president in the next 25 years. (For more on probability, see *The rise of* homo economicus.)

This sounds like – and indeed is – an extraordinary assertion, since human beings routinely disagree on all manner of topics. Yet there are technical details concealed within it. To start with, it is assumed that the two agents (let's call them Agatha and Bernard) have a 'common prior'. This means that, at some point in the past, the two made the same estimate for the probability of X. Suppose that the two meet for dinner one day at age 18, and each agree that there is a 10 per cent chance of there being life on Mars.

Then they go their separate ways, and the different experiences they have will cause them to adjust their estimates. Perhaps Agatha spends the next ten years immersed in the study of astronomy and biology, and the knowledge she acquires causes her to drop her estimate to 5 per cent. Bernard meanwhile spends the decade absorbing sensationalist newspaper accounts of alien abduction; although he treats each individual story with rational scepticism, the net result is that he increases his estimate to 35 per cent.

Ten years later the two meet again and tell each other their new estimates. Now, it is not necessary that they explain to each other how they have arrived at these conclusions. The mere fact of sharing their values will have remarkable repercussions. But it is crucial that these estimates are elevated to common knowledge between them: Agatha knows that Bernard's estimate is 35 per cent; he knows that she knows this, and she knows he knows, and so on.

With this done, the two agents will each take the other's value as a new piece of data, and each update their own estimate accordingly. They then share these new values. This process will repeat and continue as long as necessary. The extraordinary conclusion, as Robert Aumann proved, is that the two agents must eventually converge upon a single value.

Aumann's theorem embraces the notion of common knowledge. It also requires that Agatha and Bernard are perfectly rational agents who make honest and necessary updates to their predictions when the facts demand it. They do this through a particular process for taking evidence into account to arrive at probability estimates, an influential idea that goes by the name of Bayesian updating.

Taking Agatha and Bernard's originally agreed prior probability of life on Mars as 10 per cent, the probability (P) of this fact (X) being true can be expressed as $P(X) = 0.1$. Then one party is faced with some new data, D (perhaps the discovery of water on the Red Planet). Confronted with this, a rational agent needs to update its estimate to a new value expressed as $P(X \mid D)$, which is read as 'the probability of X given D'.

Now, the agent also had a prior probability for the likelihood of D itself, written $P(D)$, as well as a combined estimate for both X & D, written $P(X \& D)$. The agent's new estimate for X will be its prior estimate for X & D as a proportion of its initial estimate for D. As a formula this is

$$P(X \mid D) = \frac{P(X \& D)}{P(D)}$$

It might be that the agent judged the likelihood of water on Mars as around 20 per cent (meaning $P(D) = 0.2$) and the combined likelihood of water and life at 9 per cent (meaning $P(X \& D) = 0.09$). In this case, the discovery of water will prompt the agent to update the estimate from $P(X) = 0.1$ to

$$P(X\,|\,D) = \frac{0.09}{0.2} = 0.45$$

In other words, 45 per cent.

The illogicality of trade

Do humans actually work in this rigorously logical way? This is a question philosophers and psychologists have vigorously debated. Contra Aumann, human beings exhibit a seemingly boundless capacity for disagreement, even when in full and shared possession of all the available information. And Aumann's Agreement Theorem has thrown up some strange consequences, not least a model of economic behaviour that, pursued logically, shatters the rationale for trade relations.

In modern finance, where people trade commodities speculatively rather than just buying what they need (see *Tulip bubbles and hedge funds*), a trade requires that both buyer and seller expect to profit from the deal. But if the buyer gets a good deal and makes money on the trade – that is, he gets the commodity for less than it is worth – it must also be true that the seller would have been better off not selling. In the context of an efficient market, applying Aumann's Agreement Theorem means that the two parties will eventually come to agree on whether or not selling is sensible, with the consequence that no deal will be done.

Although the details are subtle, the intuition behind such 'no-trade theorems' is that it is not sensible to trade with someone who has good reason to think they will profit from trading with you. Needless to say, this is some distance from how markets actually function – indeed, some might say it is entirely the reverse of a properly functioning market – suggesting that financial traders may not be the perfectly rational agents of classical economic theory.

This conclusion may not be surprising, but it is certainly a worthwhile cautionary tale. An appreciation of orders of knowledge, common knowledge and how probability should be updated in the light of new information can shed light on rational behaviour. But, as with the Emperor's new clothes, human beings are adept at putting their rationality to one side at the most inappropriate moments.

AN AVERAGE CHAPTER

Mathematical definitions
of the typical

Colloquially, we may invoke the term 'average' in a loose sense, and we all know what we mean at a certain level. We're referring to something approaching a norm, perhaps implying something applicable to a majority of people as opposed to a minority, or we might be referring to an idea of everyday, middling circumstances or expectations. We might even intend it in a faintly derogatory way, as in the 'merely average'. If an athlete is described as performing averagely, while that would be no disgrace it is probably not time to break out the champagne.

But how far do these popular notions get us? What happens when we need to put a numerical value on the 'average'? In many arenas of modern life, we may need to extract the typical value from a mountain of data. Major decisions may hinge on the value we pick (see *The rise of* homo economicus). This is particularly important in statistics and economics, where analysts search for simple ways to describe complicated scenarios. Very often, controversial funding priorities – from the vast expenditure of modern welfare states down to the hiring and firing of individual workers – are based on research and trends revolving around a definition of the 'average'. It's all around us – how often, in political and economic discourse, do we hear arguments back and forth about 'average earnings', 'average life expectancy', 'average reoffending rate', 'average achievement' in schools, and so forth.

Clearly, mathematics is fundamental here, and mathematical methods for defining the typical date back to Pythagoras and his followers in Ancient Greece. Since then a number of different ways of arriving at an 'average' have been found. Indeed, one could almost say that the umbrella designation 'average' is too vague, and instead we should be talking about midranges, medians, modes and the arithmetical and geometric mean.

Between extremes: the midrange

The quickest and easiest form of average to calculate is the midrange. The shortest adult human ever recorded is Chandra Bahadur Dangi at 54.6 centimetres (under 2 feet), while the tallest is Robert Wadlow at 272 centimetres (just under 9 feet). These two figures are all that is needed to calculate the midrange of modern humans' heights: it is simply the point midway between the two, easily calculated by adding them together and dividing by 2, producing 163.3 centimetres (a little under 5.5 feet).

The midrange is, evidently, a rough and ready measurement, and its advantages and disadvantages can be seen in this example. On the one hand, it is quick and easy to calculate. On the other, it depends entirely on the most extreme – and therefore atypical – cases.

In 1890, what seemed to be giant human arm and leg bones were unearthed in a bronze-age burial-site in Castelnau in France. They appear to come from a human around 350 centimetres (almost 11.5 feet) tall. If verified, the 'Giant of Castelnau' would certainly be a remarkable discovery – and it demonstrates the flaws of the midrange principle, since it would skew 'average' human height by increasing it to 202.3 centimetres (over 6.5 feet).

Despite its limitations, the midrange was the commonest form of average for many years. For instance, the astronomer and polymath Ptolemy wanted to know the sun's angle in the sky at noon on an average day. To come up with a value, he found its maximum angle (on the shortest day of the year) and its minimum (on the longest day). The midrange of these came out at around 23.5°, which quantifies the tilt of the Earth's axis.

Bang in the middle: the median

Another type of average, the median, resists being skewed by outliers, but its principle is equally simple to grasp. It became increasingly used in science from the later 19th century. All the data points under consideration are placed in order, and the one that sits in the middle is the median. So, if five dogs have ages of 1, 2, 4, 8 and 13 years, then the median is 4. This means that exactly 50 per cent of the other dogs are younger, and 50 per cent are older.

Although this seems a commonsensical approach to take, it is only comparatively recently that the median has been treated seriously within scientific circles. The psychologist Gustav Fechner was one of the earliest to use it, in the mid-19th century, in his analysis of how people interpret light of different wavelengths as colours. Fechner found one particular property of the median very useful: that it minimizes the distance to each data-point. To demonstrate this, we can return to the five dogs, with a median age of 4. If we now calculate the difference between each dog's age and the median, we get the following five distances: 3, 2, 0, 4, 9. Adding these up gives a total of 18.

Now, we could have performed the same procedure choosing any other number in place of the median. If we pick the number 5, for example, we get the following set of numbers: 4, 3, 1, 3, 8, which add up to 19. It is no coincidence that this total is now larger than the one we got from median. Mathematical considerations guarantee that it always will be.

The most frequent: the mode

The mode is another straightforward interpretation of 'average'. In this case, it just means the commonest or most popular result. In a class of children, if seven of them are 8 years old, fifteen are 9, and two are 10, then the modal age is 9. The mode only works well in situations where the number of possible responses is limited. For example, the mode serves to define the outcome of political elections where each voter selects from among a small number of candidates (see *Ballot box paradox*). Unlike other notions of average, the mode also works in non-numerical settings.

The statistician Allen Wallis noticed an example of the mode in the writings of the fifth-century BC historian Thucydides, as he described the Peloponnesian War between Athens and Sparta. Allies of the Athenians were under siege in Plataea and wanted to know the height of the surrounding fortifications built by the Spartans besieging them, so that they could build ladders of the correct length and escape. So they counted the layers of bricks in the fortifications. Recognizing that an individual making a single count would be liable to make a mistake, the commander instructed many people to count simultaneously. They then compared notes, and picked the commonest (or modal) answer as the correct one.

The Plataean commander's idea is an example of what is now known as the 'the wisdom of crowds', and it has borne fruit in far less life-threatening situations too. For example, in guessing the number of sweets in a large jar, any individual estimate is likely to be wrong, but experiments have suggested that asking a *large* group of people each to guess, and then calculating the average value (not necessarily the mode – perhaps the median or the arithmetic mean) may produce a reasonably accurate value, as overestimates and underestimates cancel each other out.

The modern 'average': the arithmetic mean

The most well-known average in the modern era is the mean, or more precisely the arithmetic mean. (For 'regression to the mean', see *Smoke and mirrors*.) Its principle is again straightforward, and involves adding followed by division. If we want to know the mean weight of six differently sized eggs, we add together the individual weights – say 50g, 52g, 58g, 61g, 63g and 70g – and divide by the number of eggs, in this case six, which will generate a mean weight of 59g. What we are asking for, algebraically, is a single weight a, so that

$$a + a + a + a + a + a = 354g.$$

Nowadays, when people talk about the 'average' more likely than not they are thinking of the mean. This dominance dates back to the early 17th century. Its first uses, as with the mode before it, were for minimizing error, this time in navigation at sea. In order to judge the current direction of travel, a ship's navigator would use a magnetic compass for finding north. But these early devices were prone to error, particularly in stormy seas. So the standard practice became to take multiple readings, and then use their mean as the correct value (perhaps after throwing out any readings that deviated wildly from the rest).

To our eyes it seems strange that the mean took so long to come to prominence, since it had been intensively studied by Pythagorean mathematicians 2,000 years earlier. However, the Pythagoreans had not been interested in it as a statistical tool, but for its purely mathematical properties, as well as its applications to musical theory. The most famous result demonstrated by the Pythagoreans was the relationship between the arithmetic mean and yet another important notion of average: the geometric mean.

Averagely interesting: the geometric mean

An excursion into interest rates helps us grasp the relationship of the arithmetic and geometric means. Suppose we put some money in an investment and it does very well. In the first year its value rises by 50 per cent, in the second by 10 per cent and in the third by 15 per cent. Over the three years, how do we determine the average rise in value? Neither the midrange, mode nor median definitions of 'average' is appropriate here, so we might try the arithmetic mean, adding the per centages (50 + 10 + 15) and dividing by the number of years (3), giving a result of 25 per cent.

This may *seem* straightforward, yet on closer examination

the answer doesn't look quite right. Imagine that the initial amount of money invested was £100. After the first year it would increased to £150. In the second year it would increase by 10 per cent of £150 – £15 – thus yielding a pot of £165. In the third year, it would grow by 15 per cent of £165, which is $0.15 \times 165 = £24.75$, giving a final three-year value of £189.75.

Now, if our calculation of the average rise in value according to the arithmetic mean were correct, we should expect to arrive at the same final value. A 25 per cent rise in the original £100 in the first year takes it to £125; in the second year, 25 of that (£31.25) brings the total to £156.25, and in the third year another 25 per cent increase takes the final total to £195.31. There is a discrepancy.

Our investment example differs from our box of eggs in one critical respect. With our eggs it made sense to add up all the individual weights. But adding up the percentages of money really carries little information, since percentages are fundamentally about multiplication, not addition. To increase a sum of money by 15 per cent is to say the same thing as multiplying the amount it by 1.15. So, to calculate the amount after three years, we need to multiply £100 by the three numbers in turn: $£100 \times 1.5 \times 1.1 \times 1.15 = £189.75$. In this context, expressed algebraically, the right notion of average will be a number a so that $£100 \times a \times a \times a = £189.75$. What this really means is that we need $a \times a \times a = 1.8975$.

Since there are (in our example) three years of increases in value this can be found by taking a to be the cube root of 1.8975, written as $\sqrt[3]{1.8975}$.

This produces a figure around 1.23802. Now this has the right property: an initial investment of £100 which increases by this amount (23.802 per cent) each year for three years does indeed produce the correct total: £189.75.

To generalize from our example, then, the *arithmetic* mean of *n* numbers is found by adding them all up, and then dividing by *n*. For the *geometric* mean, we instead multiply the *n* numbers together and then take the *n*th root. This final step is represented by the symbol $\sqrt[n]{}$.

In turn, this may be rearranged to read as $x = y^n$, which, expressed more fully, is:

$$x = \underbrace{y \times y \times \ldots \times y}_{n \text{ times}}$$

So, for example, $\sqrt[4]{16} = 2$, because $2^4 = 2 \times 2 \times 2 \times 2 = 16$, while $\sqrt[3]{125} = 5$.

Different questions, different answers

For economists whose job it is to gauge the rate of inflation or average price rises, the Pythagorean theorem on the difference between the two means can have substantial effects (see box opposite). The primary statistical tools here are price indices, which involve monitoring a typical 'basket of goods' each year, and analysing how its average price rises or falls. In the UK, there are two major versions of this, known as the Retail Price Index (RPI), and the Consumer Price Index (CPI). They differ in various technical ways; for instance the RPI includes the cost of housing, while the CPI does not. But the most significant *mathematical* difference between the two is that the RPI is built on the arithmetic mean, while CPI is geometric. Hence, any comparison between the two will run into the bald mathematical fact that the

CONTENDING MEANS – ARITHMETIC VERSUS GEOMETRIC

The mathematicians of the Pythagorean sect discovered a fundamental relationship between the arithmetic and geometric means: whatever the numbers, the arithmetic mean always comes out bigger than, or equal to, the geometric mean.

This relationship is easiest to grasp when there are just two numbers involved, a and b. The arithmetic mean is arrived at by the formula $(a + b) \div 2$, and the geometric mean is defined as $\sqrt{a \times b}$.

Any number, positive or negative, when multiplied by itself produces a positive answer, while the value of $\sqrt{a} - \sqrt{b}$ may be positive (if a is bigger than b) or negative (if it is smaller). But either way, it will be true that

$$\left(\sqrt{a} - \sqrt{b}\right)^2 \geqslant 0$$

(the '$\geqslant$' meaning 'greater than or equal to'). A little algebraic manipulation turns this into

$$a - 2\sqrt{ab} + b \geqslant 0$$

Rearranging this further produces

$$a + b \geqslant 2\sqrt{ab}$$

With a final tidy-up, it shows that the arithmetic mean cannot take a value less than the geometric mean, no matter what the values of a and b:

$$\frac{a + b}{2} \geqslant \sqrt{ab}.$$

What is more, the 'equal to' aspect holds only when the two numbers are themselves equal: $a = b$.

arithmetic mean of a set of numbers always exceeds the geometric mean. Any sort of expenditure or investment linked to one or other of the price indices will reflect the geometric/arithmetic difference.

Most of us would prefer our pension payouts to be tied to the RPI; but if we owned a company, we would doubtless prefer to limit our wage bill by deciding employees' salary rises according to the CPI. Governments and organizations, if they switch from one to the other to arrive at a particular statistic, risk close and questioning scrutiny as to their motives (assuming anyone notices).

This all goes to show that, with averages, as with so much of science (and indeed life in general), it is not simply a matter of finding the right answers, but of asking the right questions. The truth is that there is no single 'average' of a set of data. So, before picking the best average to use, we need to consider what we want from this number, what is relevant to the current scenario, and what is not. We have to consider how we will use it, and how others will interpret it. In general, mathematics can tell us a great deal about the world, but one must not apply formulas blindly, without giving due consideration to the context in which they are being used.

ATOMIC NETWORKS AND CHEMICAL TREES

Chemical graph theory and the make-up of molecules

Not so long ago, if someone mentioned a 'network' they might be referring to a rail system, perhaps a circle of business acquaintances, or even a spy ring. For the Facebook generation, the first definition to spring to mind might be an online social networking program or a computer network. The concept of a network has pervaded science and society; we will meet many types of network in this book (see *Our electronic friends* and *Search-engine society*).

But it is not only modern technology that invokes this concept. Networks, or arrangements of linked nodes, are also of interest to chemists and physicists as they turn their microscopes on individual molecules and the atoms within in order to explore the nature of the material world. Within a molecule, it is the particular configuration of its atoms that determines its chemical properties. Since the same atoms, organized differently, alter the properties of the substance, understanding atomic shapes is at the heart of molecular science.

By calling water 'H_2O', we refer to the fact that a single molecule is composed of two hydrogen (H) atoms and one oxygen (O). In

1824, Justus von Liebig discovered a chemical – fulminic acid – whose molecules contained one each of the atoms hydrogen, carbon, nitrogen and oxygen, yielding the chemical formula HCNO. But the following year his friend Friedrich Wöhler discovered another chemical – isocyanic acid – with the same atoms, represented by the formula HNCO. Experimentation quickly established that the two really were different: when fulminic acid reacted with silver, the result was explosive; not so for isocyanic acid.

In fact, these were the first examples of isomers – chemicals containing exactly the same atoms, but in different arrangements. Since then, we know that there are still other possible permutations of the same four atoms: cyanic acid is HOCN, while isofulminic acid is HONC.

A mathematical forest

It was a chemically inclined English mathematician, Arthur Cayley, who first addressed the mathematics of this scenario in the 1870s. Today, chemical graph theory, as it is known, is a burgeoning field, which benefits from the twin approaches of viewing a molecule chemically and as a mathematical network. ('Graph' and 'network' are really synonymous terms.) What is more, chemical data, such as the melting and boiling points of a chemical, can often be predicted from knowledge of the geometrical structure of the molecule.

Appropriating some language from the natural world, Cayley explored the potential of trees – very particular types of mathematical networks. His investigations went on to form the mathematical heart of chemical analysis, and to feature applications from internet search engines (see *Search-engine society*) to the highly complex creation of timetables (see *Teacher troubles*).

A network is formed whenever a collection of nodes is joined together with straight edges. In applications within chemistry, the network's nodes represent the atoms within a molecule, while an edge between two nodes indicates a chemical bond. Every molecule can be represented in this way, and there is one thing that we can say about the resulting network, namely that it is *connected*: from any node a path can lead to any other node.

In contrast to an ordinary network, a tree is defined by the fact that (like its horticultural equivalent) while there can be branches and stems there can be no closed loops. (A loop of four nodes would be A, B, C and D, where A is connected to B, B to C, C to D, and D back to A.) Trees are among the most elementary objects in science, and can be found lurking in all sorts of settings. But, as Cayley would discover, this simplicity is highly deceptive and trees – basic though they seem – continue to perplex mathematicians today.

The first task Cayley set himself was to answer the question: how many *different* trees exist once the number of nodes is fixed? The answer depends closely on how we phrase the question, and the critical difference is between labelled and unlabelled trees. For instance, suppose there are three nodes, A, B and C. Do we count the tree labelled A–B–C as being the same as the tree labelled B–A–C? They certainly have the same shape, so if shape is all we are interested in, we should indeed count these two as the same. On the other hand, there is a difference: B is in the centre of the first tree, while A is in the middle of the second.

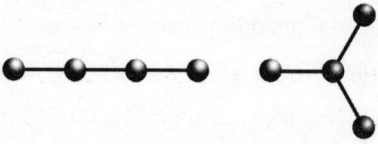

The two possible unlabelled trees with four nodes.

Cayley began by analysing labelled trees, differentating even between trees of otherwise the same shape. In chemical terms, this matters when each node represents an atom of a different element, as is the case with cyanic acid. A chemist's version of the mathematical question would be: What is the maximum number of different possible isomers of a (non-cyclic) molecule when all the atoms are different?

The number of trees with a single node is self-evidently one. Similarly, there is only one way to connect two nodes, so the number of 2-node trees is also one (we count A–B as being the same as B–A, since rotating the tree by 180° is only a superficial change). Once we reach three nodes, however, there are three labelled possibilities: A–B–C, B–C–A and C–A–B.

At four nodes, something new happens: a tree may come in two fundamentally different shapes. For trees arranged in a single row, there are 12 possibilities, beginning with A–B–C–D, then B–A–C–D, etc. In addition there are four other possibilities, where one node is placed in the middle, and the other three radiate away, their edges like the spokes of a wheel. In total, then, there are 16 labelled trees with four nodes.

Cayley contemplated the number of labelled trees as the nodes increase. This sequence begins 1, 1, 3, 16, ... but then? He came up with a beautifully simple mathematical formula. If we call T_n the total number of labelled trees with n nodes, then:

$$T_n = n^{n-2}$$

That is to say, it is the number of nodes to the power of that *same number minus 2*. So, the number of labelled trees on 5 nodes is $5^3 = 125$, and the number on 6 nodes is $6^4 = 1,296$. Having such an explicit formula is a valuable shortcut, since the numbers become mind-boggling quickly, and once we reach 50 nodes, the number of possible trees they can generate – at 50^{48} – is around the number of atoms in the known universe.

Meet the alkane family – and their relatives

Back in the chemical world, the formula C_4H_{10} represents the four carbon (C) atoms and ten hydrogen (H) atoms of the gas butane, which is one of the chemicals called 'alkanes', all composed of carbon and hydrogen. Alkanes form major ingredients of natural gas and crude oil, and thus have been hugely important as fuels, as well as in the production of pesticides and drugs, among numerous other industrial applications.

The formula C_4H_{10} does not, of course, tell us how the carbon and hydrogen atoms are connected, and in fact butane too comes in two isomers. The commonest, n-butane (the 'n' standing for 'normal') has the four carbon atoms joined in a row, while isobutane has one carbon atom in the middle and three others branching off it. The two butanes possess different properties: n-butane becomes liquid when cooled to around 0° Celsius, while isobutane does so at $-12°$ Celsius.

The first of the alkane family is methane (CH_4), the major ingredient in natural gas, while the next two are ethane (C_2H_6) and propane (C_3H_8). None of these have any isomers, but they do illustrate another important compositional principle: if an alkane

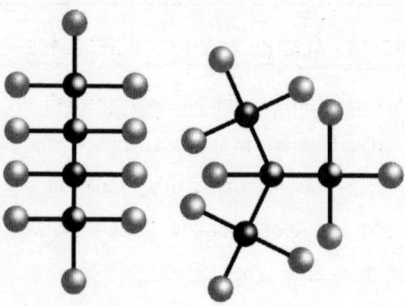

The structure of n-butane on the left and isobutane on the right.

contains n carbon atoms, then it must proportionately contain $(2 \times n) + 2$ hydrogen atoms. This makes it possible to derive, as the mathematician William Clifford proved, a generic formula – C_nH_{2n+2} – to define the whole family of alkanes (see opposite).

So far, the alkanes mentioned have all been gases; but beyond butane comes pentane, a volatile liquid often used as a solvent (for example as paint thinner), whose chemical formula is C_5H_{12}. In this case, there are three possible isomers. The standard version, n-pentane, has its five carbon atoms in a line, while isopentane has a branch coming off the main line, and neopentane has one carbon atom in the middle of a 'ring' of four others.

There is no cut-off to the value of n in the alkane formula. Kerosene contains alkanes up to *hexadecane* ($C_{16}H_{34}$). Beyond this point, alkanes are solid at room temperature. From $C_{20}H_{42}$, alkanes form paraffin wax, and are used for making candles among other things (although the name 'wax' is technically a misnomer). Alkanes past $C_{30}H_{62}$ occur in asphalt, which is commonly used in the construction industry, while those with many thousands of carbon atoms have been synthesized in the laboratory.

DEFINING THE ALKANE FAMILY – CLIFFORD'S PROOF

In 1875 William Clifford mathematically proved the validity of the alkane formula. It is worth a closer look at his elegant proof. Supposing an alkane has a formula C_nH_m, meaning that there are n carbon atoms and m hydrogen atoms, Clifford showed that it must be the case that $m = 2n + 2$.

The fundamental chemical fact is that in alkanes each carbon atom is joined to exactly four other atoms (because the outermost shell of a carbon atom contains just four electrons, which are its means of linking with others); each hydrogen atom is connected to just one other atom. If we add up all these connections for all the n carbon atoms and m hydrogen atoms, we arrive at a number expressed as $4n + m$. What is this? It is essentially the number of edges within the network, except that each edge has been counted twice – once from each end.

Now, a fundemental fact about trees is that that the number of edges connecting the nodes must be one fewer than the number of nodes, a number that can be expressed as $n + m - 1$. Since we have just said that twice the number of connections is $4n + m$, it should be the case that doubling the number of edges (so turning $n + m - 1$ into $2n + 2m - 2$) ought to deliver the same result. That equivalence can therefore be expressed as:

$$2n + 2m - 2 = 4n + m$$

Simplifying this formula algebraically gives us $m = 2n + 2$ for the proportion of m hydrogen atoms to n carbon atoms. This is the fundamental relationship between the hydrogen and carbon atoms within any alkane.

But how many isomers does each alkane have? Methane and ethane come in only one form. Butane has two isomers, and pentane three. The next alkane, hexane (C_6H_{14}), has five isomers, while heptane (C_7H_{16}) has nine: both these alkanes are primary ingredients in liquid petroleum. Octane (C_8H_{18}), meanwhile, has eighteen isomers.

Again, Arthur Cayley sought to find the pattern, and this time he had to contend with the category of 'unlabelled' trees.

The continuing mystery of unlabelled trees

The structure of an alkane molecule is fully determined by its arrangement of carbon atoms, and the hydrogen atoms stuck around the outside may be ignored. Since all the nodes in the resulting tree represent carbon atoms, there's no need to distinguish between them – they are therefore 'unlabelled', and it is just the tree's overall shape that matters.

Although Cayley had found an elegant formula for the number of labelled trees, the story for unlabelled trees is shockingly different; indeed, it is unresolved. At present, there exists no expression for the number of unlabelled trees, something of an embarrassing failure. What we do know are the first few numbers in the sequence: 1, 1, 1, 1, 2, 3, 6, 11, 23, 47, 106, 235, 551, 1,301, 3,159, 7,741, 19,320, 48,629, etc.

The best result along these lines is an approximate formula discovered in 1948 by Richard Otter. He showed that the number of trees will always be roughly $a \times n^{-\frac{5}{2}} \times b^n$ where a is around 0.5349 and b is around 2.9557.

Chemistry provides an additional subtlety here, in that for alkanes – as mentioned – one carbon atom can be joined to, at most, four other atoms. This means that trees representing alkanes are limited, in that each can have at most four edges coming off each node. Mathematicians call these 'trees of valency 4'. This limitation makes the mathematics easier to analyse, though still far from straightforward.

Despite this lack of resolution, Cayley's work on unlabelled trees did lead to the following analysis of the numbers of alkane isomers relative to the number of carbon atoms.

Carbon atoms	1	2	3	4	5	6	7	8	9	10	11	12
Isomers	1	1	1	2	3	5	9	18	35	75	159	355

Carbon atoms	13	14	15	16	17	18	19	20
Isomers	802	1,858	4,347	10,359	24,894	60,523	148,284	366,319

This table incorporates some extensions and corrections made by later thinkers, including the mathematician George Pólya in the 1930s. It is the most famous success to date in the subject of chemical graph theory, a field that continues to inspire chemists and mathematicians alike.

ALL-CONQUERING ALGORITHMS

The backbone of the
computer age

In the 21st century, even people of advancing years and little technological know-how are likely to have come across the term 'algorithm'. This word has escaped from the laboratories of software companies and the bedrooms of geeks to become a term that writers and commentators now take for granted.

Put simply, an algorithm is a to-do list: a precise set of instructions for completing a task. The theory of algorithms is arguably the single most important scientific development of the 20th century. Today, the word is synonymous with computer programs, and it is from the theory of algorithms that the stored-program computer emerged in the mid-20th century.

But the origins of algorithms long predate the dawn of the Information Age. In fact, we have to take a trip back to ninth-century Persia, during the Golden Age of intellectual efflorescence in the Middle East.

From quadratic quandaries to Turing machines

Less well known than its basic meaning is that the word 'algorithm' is eponymous, deriving from the Persian scholar

Ja'far Muhammad ibn Musa al-Khwarizmi – whose name was approximated in Latin as 'Algoritmi'. Around the year AD 820 he made breakthroughs in the study of equations. In particular, al-Khwarizmi is responsible for one of mathematics' most famous (and – by millions of schoolchildren – hated) discoveries: the formula for solving quadratic equations. The essence of a quadratic equation is that it describes an unknown number (x) in terms of its square $(x^2 = x \times x)$.

The fundamental mathematical problem al-Khwarizmi tackled was: could a value of x be found for a quadratic equation – say $x^2 + 3x + 2 = 0$ – that made the equation true? One approach is trial and error, with the obvious drawbacks, but al-Khwarizmi provided a much more efficient procedure. Given an equation involving three other numbers, $ax^2 + bx + c = 0$, he showed how to find its two solutions precisely via what passed down to generations of mathematicians (and, yes, schoolchildren) as the quadratic formula:

$$x = \frac{-b \pm \sqrt{b^2 - 4ac}}{2a}$$

(It has *two* solutions as indicated by the 'plus or minus' alternative symbols after $-b$; in the case of our example, the answers would be -1 and -2). Al-Khwarizmi did not have the benefits of concise mathematical notation, so he wrote his method in prose; to paraphrase, 'First multiply the number b by itself. Then subtract four times the multiple of a by c, and take the square root of the result. For the first solution, subtract b from the square root, before dividing by a doubled. For the second solution, subtract b from the negative square root, and again divide by a twice.

It would be equally accurate to refer to the quadratic formula as 'the quadratic algorithm'. Al-Khwarizmi's procedure had successfully taken the need for inventiveness, or even understanding, out of the subject, so that to solve any quadratic equation all that was needed was to blindly follow the steps.

While blind allegiance may not sound like a good thing in science, procedures as precise as al-Khwarizmi's would later become highly sought after for the purpose of automating calculation. It is this precision which distinguishes a genuine algorithm from a more fanciful to-do list such as 'First, cure all known diseases; second, fly to Alpha Centauri'.

In the 1930s, it was Alan Turing's idea – prior to his heroic code-breaking activities at Bletchley Park – to deconstruct instructions to such an extent that they could be implemented automatically, with no human involvement at all. The so-called Turing machines that would do this were theoretical devices, never intended as physical appliances. They were mechanical symbol-manipulators, operating on long pieces of tape, populated with symbols from some alphabet (though, in fact, just two characters are sufficient: 0 and 1). They could also switch between a finite collection of internal settings. The idea was that the device would creep along its tape, reading the symbols in turn, deleting or rewriting them, and changing its internal settings, according to simple fixed laws such as 'When you see 1, delete it, rewrite with 0, and move forward; when you see a 0, stop.' The potential of these strange contraptions is by no means obvious. But with a little playing around, it is not too difficult to design Turing machines capable of carrying out simple tasks, including basic arithmetic. In time, we would come to see Turing machines as the first proof of concept for the modern digital computer. But this technological advance was not at the forefront of Turing's mind when he came up with the idea; as with his contemporary Alonzo Church, he was interested in primarily mathematical questions – besides simple arithmetic and quadratic equations, which mathematical tasks could be solved by algorithm? And even more significantly, which could not?

With ingenuity, the range of tasks Turing machines could accomplish was revealed to be immense. In fact, through the combined work of Turing, Church and several other logicians, it later became clear that any question that could be answered by any algorithmic or automated process must also be answerable by some suitable Turing machine. In other words, Turing machines perfectly captured the important notion of computability.

Infinitely powerful?

The biggest question Turing and Church addressed was whether algorithms were powerful enough to take over mathematics entirely, eliminating all need for human creativity. Perhaps every mathematical task could be reduced to an algorithm? The eminent German mathematician, David Hilbert, considered this question in the 1920s.

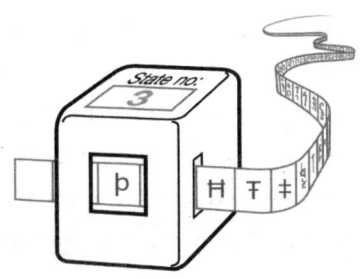

An artist's impression of a Turing machine reading and writing symbols on its long tape.

In the 18th century, an earlier German mathematician, Christian Goldbach, conjectured that every even number after 2 could be written as two prime numbers added together, so $12 = 5 + 7$, etc. It is a simple matter to write a Goldbach-checker computer program, which runs through the even numbers, checking each in turn, stopping when it finds one for which the statement is false. Indeed, at the time of writing Goldbach's conjecture has been verified up to 4,000,000,000,000,000,000,000, 000,000,000,000. If Goldbach's conjecture is *true*, this program will run forever. But if it's false, the program will halt. So if there were a device – we might as well call it an oracle – that could predict whether any algorithm would run forever or eventually halt, it would be a computational weapon of incalculable power: it would immediately settle Goldbach's conjecture, along with a slew of other major unsolved mathematical questions.

Alas, when in 1936 Turing tackled this so-called 'halting problem', he showed that no such algorithm could exist. No Turing machine (and therefore, we may be confident, no other process) would ever be able to do the job for *every* possible algorithm, a fact for which Turing provided a very ingenious proof (see opposite). And so both Turing and Alonzo Church separately realized that some problems would remain uncomputable, with the halting problem as one of the earliest to be classified as such.

Since Turing's time, it has become clear that uncomputable problems permeate mathematics. The elastic field of topology provides some good examples (see *The hole story*). Here, two shapes are considered to be 'the same' if one can be pulled or stretched (however extremely) into the form of the other. However, one is not allowed to cut the shape or 'glue' any

THE HALTING PROBLEM – TURING'S INGENIOUS SOLUTION

The halting problem, which Alan Turing grappled with, tackles the question of whether an algorithm *A*, when run on some input data *I*, eventually halts or not.

To do this, we imagine an oracle, *O*, which can always know the answer. If we then tell it the code for algorithm *A* and the input *I*, it will produce an answer: *O* will print 'halts' if *A* stops on input *I*, or 'loops' if it runs forever.

At this point, things become a little strange. Turing had the idea of making algorithm *A* work on its *own* source code. However odd this seems, *O* should still do its job by printing 'halts' if *A* stops on input *A* or 'loops' if it does not.

From here Turing concocted a paradoxical new program, *P*. When fed an input *A*, the program *P* looks at how *O* reacts to *A* and then does one of two things. If *O* outputs 'loops', then *P* prints the numbers 1, 2, 3 and stops; but if *O* outputs 'halts' then *P* begins counting 1, 2, 3, 4, 5, 6, … and never finishes.

In other words, if *A* halts on input *A* then *P* loops forever, and vice versa. In a final twist, *P* is also fed its own code. In this case does *P* count up to three or keep going forever? If the former, then by definition it must be that *O* outputs 'loops' – but that only happens if *P* never halts on input *P*.

Conversely, when P looks at its own source code, it will only count forever if *O* outputs 'halts', which means that *P* halts on input *P*. The upshot is a contradiction: *P* halts on input *P* exactly if it does not. The only resolution is that the oracle program *O* cannot exist, making the halting problem the mother of all uncomputable problems.

sections of it together. On this basis some surprising conclusions emerge – for example, that a teapot and a pair of handcuffs may be the same, but a pancake and a doughnut are different. An obvious question is whether there is an algorithm that can determine whether *any* pair of shapes are the same, just as al-Khwarizmi's formula would solve any quadratic equation? Remarkably, it turns out that no Turing machine will ever be able to do this job, and therefore we can be fairly confident that no other means of computation can do so either. That is not to say that these questions have no solution; on the contrary, any two shapes will either be the same or not. But the point is that there is no single method that can reliably decide which.

From counting to computing

Lest we become too negative about algorithms' limitations, we need to remind ourselves of the modern computer's transformational effect on the world. That potency comes from the *programmable* nature of a computer, the essential characteristic that differentiates it from the pocket calculator.

The theoretical underpinning of today's programmable computer has its roots in Turing's work on the universal Turing machine. The possibility of such a device was one of Turing's greatest discoveries, though not as hard to prove as one might expect. While the range of all possible Turing machines is immense, a universal Turing machine is able to emulate every other when provided with the correct instructions, in the form of the appropriate sequence of symbols on its ticker tape. Quite simply, anything that can be computed by any mechanical means whatsoever, can be computed by a universal Turing machine.

Machines that count, of course, have a long lineage. A notable moment came with Gottfried Leibniz's 'stepped reckoner' in 1672, a gadget capable of addition, subtraction, multiplication and division. In the mid-19th century, Charles Babbage's Difference Engine was designed to go much further. In the early 20th century progress continued, but there was a sense that the final goal was never quite clear. It was with Alan Turing's description of a universal machine that the standard for computation was finally set, and groups around the world now knew what they had to aim for: a physical device with the same capabilities.

The hurdles were huge. The Turing machine was immensely important as a theoretical breakthrough, but as a practical design it was a non-starter: it could only communicate with the user through great lengths of ticker tape (which would inevitably get torn or tangled) and it had to print, scan and interpret printed symbols: a messy, slow, expensive and error-prone process. All in all, it seemed a deeply unpromising blueprint. Yet the promise that it held out, of universal computation, was a prize worth fighting for.

One of the earliest functioning machines was the Electronic Numerical Integrator and Computer (ENIAC), built by the US Ballistics Research laboratory in Maryland and first switched on in 1946. It was a monster, large enough to fill a small apartment, weighing over 27 tons and incorporating more than 17,000 vacuum tubes. In place of reams of ticker tape, it received its data on punched cards and stored the information internally using 'flip-flops': electronic switches that could adopt one of two positions. Programming it was a laborious business, and it involved weeks of plugging cables and flicking switches into the correct configuration. All the same, the 'giant brain', as it became known, did work. Once properly set up, it could perform up to 5,000 calculations per second. Nevertheless it left a great deal of potential for improvement.

The universal Turing machine provided a hint of one important principle of computer design. Every algorithm or Turing machine receives an input on which to run. For example, to solve a quadratic equation, we need inputs to tell us the values of a, b and c. On top of this, a universal Turing machine receives a special input, describing the algorithm to be performed. We might describe these as the data and program respectively. But it is striking that a universal Turing machine does not distinguish between the two: both are just symbols on the tape. The ENIAC *did* make a distinction, so while it read data from punched cards, the program was stored as hardware, in the arrangement of cables and switches that its long-suffering operators had to set up. As the mathematician and computer pioneer John von Neumann recognized, this complication was unnecessary: Turing had showed that it should be possible to store and manipulate programs *as* data.

This dream of the stored-program computer was made a reality in subsequent machines such as the Electronic Discrete Variable Automatic Computer (EDVAC). Switched on in 1951, it used its memory much as a Turing machine used its tape, both for reading and manipulating data, and for storing programs to be run. In Europe, Konrad Zuse had had a similar insight as early as 1936. His 1941 machine, the *Z3*, anticipated many of the US developments, but went largely unnoticed at the time because of Nazi Germany's political isolation.

The EDVAC was an improvement on the ENIAC in other regards too: while the old machine had worked on the usual base-ten (decimal) number system, EDVAC used base-two (binary system) to store data. So while the ENIAC had to find ways to store and manipulate the digits 0–9, the ENIAC only needed to deal with 0 and 1. These binary digits (which we better know as 'bits'; see *Avoiding bad language*) can easily be stored as the position of a single switch – or in the magnetization of small units in today's hard drives.

The computer – more specifically, the stored-program, binary computer – is surely one of the greatest inventions in our species' history, with an unmeasurable influence on today's world. And the computer has many parents. Charles Babbage has popularly been called 'father of the computer' for his Analytical Engine, successor to the Difference Engine, while Ada Lovelace, who worked with the device, is sometimes known as the mother of computer programming. With good reason Alan Turing is lauded as the father of the 20th-century's computer revolution. But the subject's underpinnings lie in the seemingly dry and impractical considerations of mathematical logic, and the questioning mind of a ninth-century Persian.

GETTING A PROPER PERSPECTIVE

Projective geometry and the world of art

When designing a railway, any engineer worth his salt will know that there is one fundamental rule that must be obeyed no matter what: the two rails must always be the same distance apart – parallel. Any deviation from this principle, beyond the narrow bounds of tolerance, will end in derailment and disaster. But an artist's representation of those rails on flat canvas or a piece of paper is quite a different matter. Unless the view is an aerial one, the painter will have to use his skills to create an illusion of parallel lines. If the rails are meant to extend into the distance, he will depict them becoming ever closer together as they recede, before finally meeting at a point – the vanishing point – on the horizon. In this way a sense of three-dimensional perspective is created.

Around the world, some schools of art have eschewed perspective, and young children's drawings certainly play fast and loose with perspective and scale. Medieval art stretched to approximations of perspective, but it was with the Renaissance that Western art really began the careful analysis of vanishing points, allowing artists to render a realistic sense of depth. In the words of critic E.H. Gombrich, in his classic history *The Story of Art*, 'It was Brunelleschi who gave artists the mathematical means of solving this problem; and the excitement which this caused among his painter friends must have been immense.' This pioneering Florentine primarily excelled in architecture rather than paint, designing Florence's celebrated cathedral; but his experiments with perspective proved profoundly influential, especially after their elaboration by fellow Italian Alberti in the treatise *De pictura*.

Behind the invention of the 15th-century Renaissance masters through to the 19th-century realist painters, and even the imaging techniques of the computer age, lie the mathematical foundations of perspective: projective geometry.

Horizons and dualities

The earliest painters to experiment with perspective tended to use a single vanishing point, plumb in the centre of the canvas. Towards this spot, avenues of trees or rows of houses could recede and disappear, drawing in the eye. But there was no need to be so limited. Artists quickly learned that multiple vanishing points were possible. If (to continue our example) the depicted railway contains a fork, the picture might show two sets of tracks disappearing into the distance in different directions, producing two vanishing points.

So, one might ask, how many vanishing points are possible? The answer is boundless: one for every direction along which a set of rail tracks might run. This collection of all possible vanishing points takes on the form of a line, known as the 'vanishing line', or in everyday parlance the horizon.

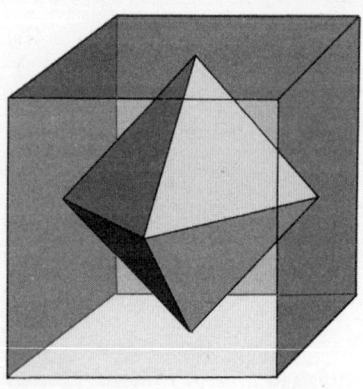

The cube and the octahedron are dual shapes.

As mathematicians investigated the properties of perspective, they encountered the fundamental fact of duality – that many geometrical objects and theorems arise in pairs. This useful principle first manifested in the theory of three-dimensional solids. For example, imagine a cube; we draw a dot at the centre of each face, and then connect the dots on adjacent faces with straight lines. If we then discard the original shape, and expose the skeletal shape just created, a new three-dimensional polyhedron is revealed: an octahedron. Mathematicians regard these two solids as duals of each other: the corners of a cube represent the faces of an octahedron, and vice versa. Repeating the procedure for the octahedron will give us back a cube. This relationship can be seen in the two shapes' vital statistics: while the cube has six faces and eight corners, the octahedron has eight faces and six corners. What is more, four faces of the octahedron meet at every corner, which reflects the fact that the cube has faces with four edges (i.e. squares). Going the other way, on a cube three faces meet at each corner. It is no coincidence that the octahedron is built from triangles.

Similarly, the dodecahedron with its 20 corners and 12 pentagonal faces, which meet in threes, is dual to the icosahedron, with 12 corners and 20 triangular faces, meeting in faces of five edges.

These four highly symmetrical shapes are known as the Platonic solids, meaning that their faces are all identical, regular polygons. The fifth and final Platonic solid is the tetrahedron, consisting of four triangular faces, meeting in threes, with four corners in total. Interestingly, this shape is self-dual; interchanging the faces and corners produces another tetrahedron.

Duality appears throughout geometry, including in the first geometrical scenario to be studied in depth: the infinite flat, two-dimensional plane, famously investigated by Euclid in his book *Elements* around 300 BC. The initial directions of enquiry here involved the very simplest objects: individual points and straight lines.

One of Euclid's axioms for the plane states that 'Given any two points on the plane, there is exactly one straight line that can be drawn to connect them.' And just as two points may be connected to produce a straight line, so two lines intersect to produce a point: the dual statement to Euclid's axiom is the law stating 'Any two straight lines meet at exactly one point.' This duality was exploited over time to prove many more elaborate geometrical theorems, but there remained a niggling problem: it wasn't entirely true. Parallel lines, inconveniently, violate the dual law, since they do not intersect anywhere at all.

Points at infinity – the projective world

This is where the world of art re-enters the discussion. To address this geometrical obstacle, in the 17th century Girard Desargues noticed that there was a context in which parallel lines do meet: the artist's canvas. His grand idea was to treat the Euclidean plane in the same manner as the ground within a painting. In addition to all the usual possible locations on the plane, he added in an extra collection of vanishing points, or as they are known in a mathematical setting, points at infinity.

This expanded space is known as the *projective plane*. It consists of the ordinary plane, along with an extra straight line where the points at infinity live. Technically, it is a little more difficult to explore than Euclid's original plane. The pay-off is that the duality between points and lines now works perfectly: every pair of points can be connected by exactly one line, and every pair of lines meets at a unique point. In this new setting, parallel lines actually cease to exist.

This development led to many beautiful geometrical results, including a refinement of one long-revered ancient theorem and the framing of a new one.

Although first framed long before the discovery of projective geometry, the theorem proved by Pappus of Alexandria in the fourth century AD was one of the best known results in the subject. Take any two straight lines, and on each pick any three points. Call these A, B, C and a, b, c. Then, draw six new straight lines, the first joining A to b & c, the next pair joining B to a & c, and the last connecting C to a & b. Now we focus on three special places: the first is where the line Ab meets the line aB. The second is where Bc intersects bC, and the third where Ac

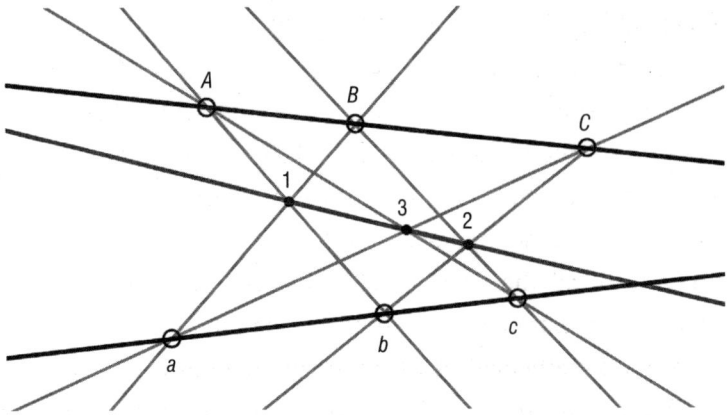

Pappus's theorem asserts that the points 1, 2, and 3 must lie on a straight line.

meets *aC*. Pappus's result was the beautiful and unexpected fact that these three points must also lie on a straight line.

The projective world provides a solution to the one awkward exception to Pappus's rule, when the lines *Ab* and *aB* are parallel. The problem goes away, because in projective geometry we can be sure these lines will intersect somewhere.

Returning to artistic matters, in a realistic depiction of a house, say, certain features should appear nearer to the viewer, and correspondingly larger, than others intended to appear more distant. We are all familiar with the effect, which artists know as *foreshortening*. To make the illusion convincing is a tough trick.

Geometrically, the requirement is that the real house is *in perspective* with the painted house, meaning that if the real house and the painting were lined up, a ray of light from each part of the house should pass through the corresponding place in the painting, en route to the viewer's eye. But how to gauge that? Using the new subject of projective geometry, Girard

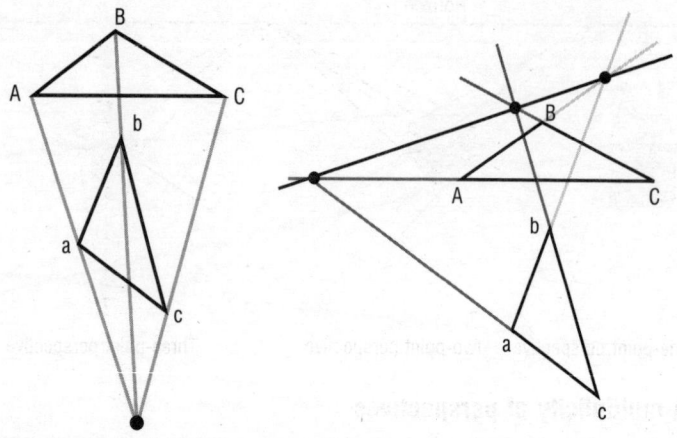

Desargues's theorem provided a convenient criterion for determining whether or not two triangles are in perspective.

Desargues found an unexpected criterion to determine whether two triangles were in perspective. If we imagine two triangles that are in perspective, whose corners are *ABC* and *abc*, then if we connect *A* to *a*, but then extend the line onwards – and do the same for *B* & *b* and *C* & *c* – the three new lines will all converge upon a single point, which in artistic terms equates to the viewer's eye.

Desargues now proved that these two triangles are in perspective precisely if a second, seemingly very different criterion holds. If we take the corresponding *sides* of the two triangles, *AB* & *ab*, and extend them in both directions, projective geometry guarantees that they will cross somewhere. If we then do the same thing for the other pairs of sides, *AC* & *ac* and *BC* & *bc*, we will obtain three new points. Desargues's theorem states that these three points lie in a straight line precisely if the two triangles are in perspective.

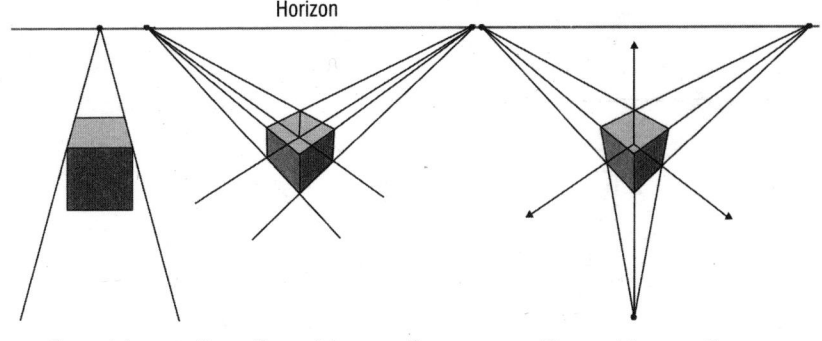

One-point perspective   Two-point perspective   Three-point perspective

A multiplicity of perspectives

Desargues's realization was a major moment in understanding the geometry of perspective, and opened up an array of new techniques for visual artists, equipping them to handle an increase in the number of perspectives.

Suppose an artist is sketching a building, aiming to impart a sense of depth. The left and right hand walls, though parallel in real life, may be depicted converging towards a vanishing point far behind the house. Similarly, the front and back walls may be drawn so that they are angled towards a second vanishing point, off to the right of the picture. Both of these vanishing points will lie on the vanishing line or horizon. However, if the artist has decided to view the scene slightly from above, it may also be that all the walls converge towards a third vanishing point, deep below the ground.

This technique of three-point perspective produces very striking illusions of three-dimensionality. It also goes to show that the power of projective geometry is not limited to the two-dimensional plane. Viewed mathematically, the whole of three-dimensional space itself may also be expanded into a projective space: in three-dimensional space there is one new point for every possible direction, which means that the collection of additional points 'at infinity' comprises not a line this time, but a two-dimensional plane.

The beauty of projective spaces in the eyes of geometers is that they avoid awkward exceptional cases such as parallel lines or planes. But in the 21st century the concept of projective space is playing a central role in technological developments such as computer vision. When a human viewer looks at a two-dimensional picture, drafted in perspective, he can automatically recreate the three-dimensional scene in his head. Technologists know this process as 'image formation', and in the digital age it has become increasingly important to understand how it works. The intention of computer vision is to design software that can do the same, and projective geometry is integral to that endeavour, with applications in, for example, robotics (see *Automata and articulation*). This is a natural development, since a programmed robot armed with a camera, looking along a pair of railway tracks, will see them meet at the vanishing point just as humans do.

Projective geometry is also involved in sophisticated attempts to understand orientation and spatial memory. For example, suppose a robot is taken to the door of a room, and shown the scene inside. Later, it is placed somewhere within the room. The challenge for the robot is to work out, from what it sees and remembers, where it is positioned in the room relative to the door. Now, it might recognize some familiar items: the clock on the mantelpiece, the green mug on the coffee-table, and so on. So the robot has a couple of two-dimensional pictures to work with and compare: its current field of vision and its earlier view. It will scan both, trying to match up points in one with points in the other. The question is: how many pairs of points does it need to match, in order to solve the problem? It is clear that one or two are not enough.

In 1997 two researchers, Olivier Faugeras and Steve Maybank, deployed subtle arguments from projective geometry to show that when five points are matched up, it is possible for a robot to deduce its new location exactly. But it is not a case of any five points. If the five are chosen unhelpfully, then the robot may be reduced to a one-in-ten chance of pinpointing its new position. Maybank also showed the surprising result that if the specified points were sufficiently inconvenient, then no number of them would serve to pin down a unique location.

From a different perspective – that of history – it has been a long but productive road from Brunelleschi's vision to today's robotic vision, a journey populated by the efforts of mathematicians, artists, computer scientists and others. Moreover, interesting episodes from the past occasionally revisit the present. Centuries ago, Byzantine iconographers employed 'reverse perspective' in their art, positioning the vanishing point not on the horizon but in the forefront of the picture, so that as the picture receded the people and objects within it became larger, not smaller. For the Orthodox Christians of the Byzantine Empire, this perhaps embodied God's all-seeing nature. The same technique has been rediscovered by modern digital artists, intrigued by its visually unusual results.

OUR PIXELLATED PLANET

The mathematics of
digital photography

We have come a long way in the world of photography since the daguerreotype and the stiffly posed Victorian studio portrait, with obligatory column and pot plant lurking in the background. Some things endure: any good photographer knows that the quality of an image is conditioned by the composition, the exposure time, and of course the nature of the subject. Every photographer wants a top-quality camera and the best lens he can afford. And photographers have always shared with the Impressionist painters an obsession with the nature and quality of light – in the end, what else does an image contain but light and colour?

But in other respects there has been a revolution. If, aesthetically, we still judge the final results of a photographer's art by traditional standards, it is very likely – almost inevitable now – that those results are being created by 21st-century digital means. This transformation from the era of darkrooms, film and chemicals to the new world of computer screens and binary information has fundamentally changed the technical process for creating a final image. Indeed, it has changed the very notion of what 'final' means: Where once there was modest airbrushing, we now go to town with Photoshop.

Central to this digital revolution are pixels – individual points that can take on a single colour, and which, in their millions, populate the modern digital photograph. How best to store and manipulate this information – and in the digital world, we need to remember that this *is* 'information' – is a profoundly mathematical question, where theoretical advances have led to great and continuing improvements in technology.

Pixels and their properties

It is widely understood that the more pixels a digital picture has, the higher its resolution. A high resolution means we can zoom into the image further before the image begins to fragment noticeably into its constituent pixels. Today's digital cameras typically produce pictures of 3–16 megapixels (1 megapixel = 1 million pixels), though cameras of up to 200 megapixels have been built. However, unless we wish to print a picture the size of a football field, the latter will not be of much use to us.

Digitally speaking, pixel density is one determinant of quality; but a second factor is the range of colours that each pixel can take on. In an image, each pixel will adopt one colour from the available options. The bare minimum is where it is either black or white. A computer will express this in binary code, assigning each pixel a single binary digit ('bit'): 0 for black or 1 for white.

Binary code also works for describing images with more colours. Some early computers, such as the famous Acorn-built BBC Micro system of the 1980s, used three bits to describe the state of a pixel, corresponding to the three basic colours:

red, green and blue. The first bit described whether the red component was turned on or off, with the second and third doing the same for green and blue. So 000 was black (all three components switched off), 100 was red, 010 green and 001 blue. Combinations of these three produced further possibilities: 110 represented a mixture of red and green – yellow – while 101 was red + blue, giving magenta, and blue + green formed cyan, 011. Combining all three basic colours gave white, 111.

This approach therefore yielded eight colours in total, because the three bits could take one of two values. It can be expressed mathematically as $8 = 2^3 = 2 \times 2 \times 2$. Of course, modern images have rather more than eight colours available. The TrueColor standard assigns each pixel a code of 24 bits, such as 10011001 00000000 00110011. As with the simple RGB system (Red Green Blue), the 24 bits are divided into three sections, this time into 3 bytes (a grouping of 8 bits), with the first byte describing the amount of red, the second the amount of green, and the final byte the amount of blue. Our specific example above creates a reddish pink. In total, the TrueColor system allows each pixel to take one of 2^{24} different colours – that is, 16,777,216 possibilities! This is likely to be enough for even our fussiest needs, since it is believed that this quantity exceeds the number of colours that the human eye can discriminate.

From these coloured pixels the picture emerges. In digital terms, what we have is a list of information describing each pixel. By way of example, we can work with a simple image comprising a square 8×8 grid of pixels. We have therefore 64 pixels in total, which we can denote as p_1, p_2 and so forth up to p_{64}. For now, we will also work in greyscale, so that each of the 64 pixels has a colour between 0 (white) and 100 (that's one hundred, representing black). We can now tag each pixel (p) with its colour from this selection, which we can call c. This means that the whole image is described by the algebraic expression:

$$c_1 p_1 + c_2 p_2 + \cdots + c_{64} p_{64}.$$

As a simple example, let's imagine a picture where every colour is mid-grey (50 on the scale of 1–100), except for the very first, which is white (0). In this case, the image would be described as:

$$0 p_1 + 50 p_2 + 50 p_3 + \cdots + 50 p_{64}$$

This image is stored by remembering the values for all the colour codes, in this case $c_1 = 0$, $c_2 = 50$, $c_3 = 50$, ..., $c_{64} = 50$.

The essence of the idea is fairly clear. But our 64-pixel greyscale image is puny in an era where 10-megapixel cameras are common. In combination with TrueColor images, these create a vastly greater demand for digital storage space. A 10-megapixel TrueColor image will eat up 30 megabytes of space, making it slow to transfer from camera to computer, potentially clogging up a hard drive, and cumbersome for email transmission.

Luckily, there are tricks to compress the image into a lighter file, with little or no visible drop in quality, and this is an area where some more serious mathematics enters the frame. It works by exploiting patterns in the data. For example, a holiday snap taken on a sunny beach may consist in large part of plain blue sky. It seems a waste to give identical instructions for every same-coloured pixel to the effect that 'p_1 is light blue', 'p_2 is light blue', 'p_3 is light blue', *ad nauseam*. This would require the code for light blue being stored millions of times over. Much better would be to indicate with *one* instruction that the top third (or whatever the area may be) of the picture is all blue, in effect saying something like 'the first 3 million pixels are all light blue'. If we enact this sort of contraction in our previous 64-pixel image of almost all grey, we could describe the entire image in two instructions:

$$c_1 = 0$$
$$c_2 = c_3 = \ldots = c_{64} = 50$$

In actual practice, standard image encodings such as JPEG format (named after its developers, the Joint Photographic Experts Group) do not take *exactly* this approach. Instead of specifying the exact colour at every point, or even within a particular region, they instead express the picture as a whole in terms of sums of *basic images*. For our 8×8 greyscale grid, there is a library of 64 such basic images, which go by the more technical name of 'discrete cosine functions'. On their own, these basic images are of limited use; but wonderful things can happen when we weight and combine them.

This can be illustrated if we return to our 8×8 grid, and call the 64 basic images $B_1, \ldots, B_{64}$. We can now weight *each one* on a scale of 0 to 100, where 0 means 'don't display it at all', and 100 means 'display it at full strength'. So the image expressed by $95B_1 + 5B_2$ means that B_1 is displayed at 95 per cent strength, overlaid with just a hint of B_2 reduced to 5 per cent strength.

The felicitous mathematical theorem here is that *any picture whatsoever* on our 8×8 grid (call it P) can be expressed by some appropriate weightings of these basic images. That is to say, when the basic images ($B_1, \ldots, B_{64}$) are weighted according

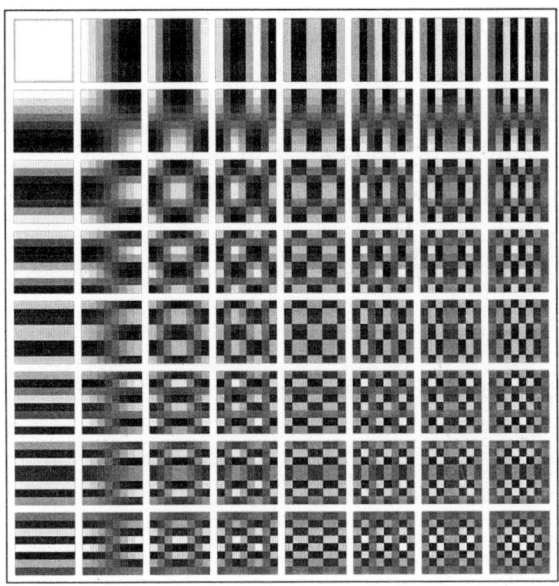

The 64 basic images for a 8 × 8 greyscale grid.

to some suitably chosen numbers, which we may call w_1, ..., w_{64}, the combined result is the picture we wanted, P. Mathematically we can express this as:

$$w_1B_1 + w_2B_2 + \cdots + w_{64}B_{64} = P$$

While elegant in its own terms, is there a saving in terms of memory usage? Seemingly not. After all, we have traded in 64 pixels, each specified by a colour between 0 and 100, for 64 basic images, each weighted with a number between 0 and 100. The overall number of pieces of data to be stored has remained the same, at 64. In terms of image management, though, there are reasons to prefer this deconstructive approach. In particular, this method corresponds better to patterns within the picture as a whole. Features such as a monochromatic sky, or a patterned wallpaper, are more likely to stand out in the data.

There is also a form of cheat that can reduce the storage requirements for the basic-image approach. If we opt for what is known as a lossy method of compression, meaning one entailing a fundamental drop in quality, it is much easier to do so it in a way that the human eye is unlikely to notice. Experience tells us that a subtle difference in the weight of a high-frequency image (bottom and right) is much harder for humans to detect than those for a low-frequency one (top and left). The implication of this observation is that we might be able to prioritize high quality where it matters, while saving space by allowing quality to drop where it will not be noticed. For instance, for the high-frequency component, instead of giving the weighting on a scale of 1 to 100, it could be more coarsely graded, on a scale of 1 to 10.

More than this, major space-savings can be made by simply forgetting all of the basic images whose weights are below some predetermined threshold, say a weight of $\frac{3}{100}$ or lower. This will drastically reduce the underlying data, without being noticeable to the human eye.

All the pictures in the world

How many possible pictures are there? It sounds like an imponderable philosophical question. But in the digital world, it is less abstract. On our monochromatic grid, for example, there are 64 pixels, and each can take a value between 0 and 100. The total number of possibilities here is an actual number – 64^{101} – in other words, more possibilities than there are atoms in the observable universe, and that just for a tiny greyscale grid! For a

10 megapixel TrueColor image, the answer comes out around a mind-bending figure of $10^{117,440,512}$. Among this gigantic number is literally every digital photograph that any human armed with a 10 megapixel camera can, or will ever, take.

Yet these potentially interesting images are hugely outnumbered by others that will *never* be taken, which we might characterize as a kind of pixel wasteland of brown sludge. Is there some way to separate the smaller number of interesting images from this visual static? If that pool of interesting pixel combinations *could* be isolated, it would be much quicker to identify any particular photograph from this useful sub-space than from the entire realm of all possibilities. It is, of course, a question that has already been tackled.

As mentioned, the basic images described above are known as 'discrete cosine functions'. In each function, the shade of grey follows what mathematicians know as a cosine wave, with the one difference that while an ordinary wave changes perfectly smoothly, the pixellated image proceeds in jumps, pixel by pixel, that is to say *discretely*. In the top images, the wave is from left to right, and in the left-hand images it runs from top to bottom. Additionally, the *frequency* of the wave varies among images, which is low at the top left, and high at the bottom right. (For more on wave forms, see *Wave-worlds*.)

However, there are other ways to break up an image. The method used in the JPEG 2000 file format (rather than the more common JPEG) involves two-dimensional 'wavelets', though the basic idea remains the same. In a 10-megapixel image, the number of weightings needed to express a picture is 10 million, rather than the 64 of our small grid. But there is a crucial observation here, which distinguishes interesting pictures from noise: for an interesting image, around 99 per cent of these weightings will be zero, or near enough to zero to be discarded. That means that only 1 per cent, or around 100,000, of these values will actually be required.

If we knew in advance which basic images those crucial 1 per cent were going to be, then the whole procedure would be much easier. It sounds impossible, but remarkably in 2004 Emmanuel Candès and Terence Tao found a way to do it. Their technique of 'compressed sensing' was extremely unexpected and counter-intuitive. Rather than measuring the picture according to the intensity of individual pixels, or the weight of wavelets, they measured it by seeing how well it matched a completely random image. In fact, they quantified the extent to which the given image resembled around 300,000 different random images (tripling the likely number required, just to be safe).

Crucially, on the random images each wavelet has a unique signature, observable in the data. From these readings, it is possible to estimate which wavelets are involved, as well as what their weightings should be. This process will not work for a completely general image, but Candès and Tao's theorem showed that it would work, with high probability, so long as not too many wavelets were involved in the original image. Since this is exactly the criterion satisfied by 'interesting' images, the process proved astonishingly effective. Proof of the concept's efficacy came in the form of a remarkable 'single-pixel camera' built by Richard Baraniuk and Kevin Kelly. With just one visual detector gauging the scene against a sequence of random images, it has been able to recreate images from their wavelet basis with an unexpected accuracy.

The advantage of compressed sensing is that it cuts by around 97 per cent the amount of information the camera needs to capture and store. This may not be especially useful for hand-held cameras; but for autonomous sensors, which remain in one place and gather information over a long period, this is a major boon.

Compressed sensing relies on the fact that interesting pictures are 'sparse' – only a small proportion of wavelets appear within the picture. Similar techniques of 'sparse representation' have been applied to other areas, including facial recognition technology. A (previously unseen) human face can be described by the extent to which it resembles other, already known, faces. But again this is likely to be sparse – the face may resemble fairly closely a few other faces (people of similar age, sex, race, and so on), but it will be broadly dissimilar to a large majority of others. This insight allows sparse representation to come in, speeding up the process of analysis and recognition.

With compressed sensing and sparse representation, the sky – pale blue or otherwise – is the limit. Further potential applications for this exciting new technology include sensors mounted on satellites and unmanned space expeditions, and it all comes from taking a modern mathematician's eye view of the venerable art of photography.

THE DYNAMIC SOLAR SYSTEM

The mathematics of
planetary motion

The astronomers of the Ancient Greek world noticed something unexpected in the night sky: among the thousands of pinpricks of light were a few that visibly moved. They dubbed them *planetes* (πλανετες), meaning 'wanderers'. The relationship between our own world, our moon, the sun, the stars of the night sky and these mysterious wandering planets has been a spur to scientific enquiry for thousands of years. It was not merely a matter of curiosity. In the seafaring age, a thorough understanding of the night sky was essential to navigation, with many thousands of lives and the fortunes of nations resting on the outcome.

For many centuries, therefore, deciphering the geometry of the heavens was arguably the single most important practical application of mathematics. It was finally placed on a solid scientific footing during the 16th and 17th centuries, through the work of some of Europe's greatest scientific minds.

The earliest cosmologists were faced with a confusing picture. As long ago as 250 BC, the Greek astronomer Aristarchus suggested a heliocentric system, one that placed the sun at its centre. The idea was not popular. People reasoned that if the earth were really travelling such great distances through space as this would suggest, we would surely notice corresponding changes in our views of the stars. Aristarchus made a significant observation to overcome this objection: the stars must be much further from us than previously believed. But in the absence of solid evidence, this was not enough to win over the sceptics.

Other models were suggested too. The Pythagorean philosopher Philolaus had earlier claimed that the earth, sun, planets and stars were all locked in orbit around a greater object, known as the 'central fire'. Opposite the earth, he argued, was a counter-earth, permanently blocked from our view.

Yet it was the geocentric model, with the earth at the centre, that became the standard view, and the version described by Ptolemy in the second century AD became the dominant cosmology for over a millennium. Ptolemy placed the earth at the centre of the universe, and the other celestial bodies in a sequence of nested spheres centred upon it: first the moon, then Mercury, Venus, the sun, Mars, Jupiter, Saturn, the fixed stars, and finally the sphere of the 'prime mover', which caused the whole system to rotate.

The advance of the heliocentrics

In the modern era it was Nicolaus Copernicus (1473–1543) who placed the sun back in the centre of the solar system, an idea that would eventually supplant the Ptolemaic consensus. Copernicus's model, described in his 1543 book *De Revolutionibus Orbium Coelestium* (*On the Revolutions of the Celestial Spheres*), went a long way towards explaining the observed movements of the heavenly bodies. Copernicus understood that the earth was a planet, just like the others. He realized too that, while the moon really did orbit the earth, the apparent motion of the sun was illusory. Its morning rising and evening setting were merely consequences of the earth's own movement, spinning on its axis, while simultaneously orbiting the sun.

Copernicus's work was subsequently backed up with experimental evidence gathered by Galileo Galilei (1564–1642), from a telescope of his own design. But as the evidence for the theory grew, so too did its controversy. Copernicus himself had been a Catholic cleric, and the Catholic Church had initially largely ignored his writings. Nevertheless, official doctrine remained fully behind the old geocentric model, and heliocentricism was deemed 'altogether contrary to Holy Scripture'. It seemed inconceivable that Man, created in God's image, would be located on a peripheral object rather than centre-stage. As is well known, Galileo's support for the Copernican theory became a source of great conflict, especially following the appearance of his *Dialogo sopra i due masimi sistemi del mundo* (*Dialogue Concerning the Two Chief World Systems*) in 1632. Galileo was tried for heresy the following year and forced to acquiesce; he lived the remainder of his life under house arrest. Nevertheless, his views continued to permeate contemporary thought, and editions of his works appeared in parts of Protestant Europe. The cosmos was finally beginning to fall into place.

While in Italy Galileo was busy establishing the basic truth of the Copernican model, the German-born Johannes Kepler (1571–1630) was already addressing the next question: if the earth and planets do indeed orbit the sun, what are the shapes of their paths? The obvious answer, accepted by both Copernicus and Galileo, was that the orbit was circular – an understandable but erroneous conclusion. The correct answer arrived in the early years of the 17th century, taking advantage of data amassed in the previous century by the Danish astronomer Tyco Brahe. Unlike Galileo, Brahe preferred to make his observations with naked eye rather than telescope. Nevertheless, over 38 years of work he amassed a table of astronomical observations more detailed than the world had ever known. Brahe intended this as support for his own 'Tychonic' model of the universe, a hybrid of the Ptolemaic and Copernican systems, in which the sun and moon orbit the earth, while the planets orbit the sun. But the real breakthrough came from Brahe's assistant, Kepler.

As with Galileo, Kepler encountered conflict with the religious authorities. A devout Christian, Kepler trained to be ordained as a Lutheran minister, but again his heliocentric beliefs posed an obstacle, flying in the face of Martin Luther's own views on the matter. Years later, Kepler would be excommunicated, causing him considerable personal pain. His dealings with the Catholic Church were no happier, and in 1600, Kepler fled his home in Catholic Austria with his family, to work with Brahe in Prague, where Brahe was working under the patronage of Emperor Rudolph II. Brahe died just a year later.

Already a supporter of Copernican theory, Kepler, as he pored over Brahe's data, noticed that the orbit of Mars, in particular, was incompatible with a circular motion. However, another related shape was a perfect match: the ellipse.

In appearance, an ellipse looks like a circle but somewhat squashed or stretched. Its mathematical definition is similar too. A circle is defined by two pieces of data – the point at its centre (C) and a specified distance from that centre, its radius (r). By definition, a circle is the set of all points that are exactly a distance r from C.

An ellipse can be defined in a similar way, but this time, the single point at the centre is replaced by two special points, the *foci*. Whether or not a point lies on the ellipse depends not on its distance to the centre, but on the result of adding together its distance to each focus. As with a circle's radius, this number is the same for all points on the ellipse.

In algebraic terms, a circle comprises the set of all points with coordinates (x, y) where x and y satisfy the equation:

$$x^2 + y^2 = r^2$$

This can be reconfigured to read:

$$\frac{x^2}{r^2} + \frac{x^2}{r^2} = 1$$

The equation of an ellipse looks rather similar to this, but it involves two numbers a and b, known as the semi-axes, in place of the single radius r. An ellipse has the equation:

$$\frac{x^2}{a^2} + \frac{y^2}{b^2} = 1$$

Here the number a (the 'major semi-axis') represents the maximum distance from the centre to a point on the ellipse, and b (the 'minor semi-axis') represents the minimum such distance.

In terms of this equation, we may define a circle to be the special case where $a = b = r$. In this case the two equations come out to be the same, and the two foci merge into one at the centre. This bit of algebraic trickery justifies the perception of an ellipse as a slightly stretched circle. It also means that many of the techniques successfully brought to bear on circles can be applied to ellipses.

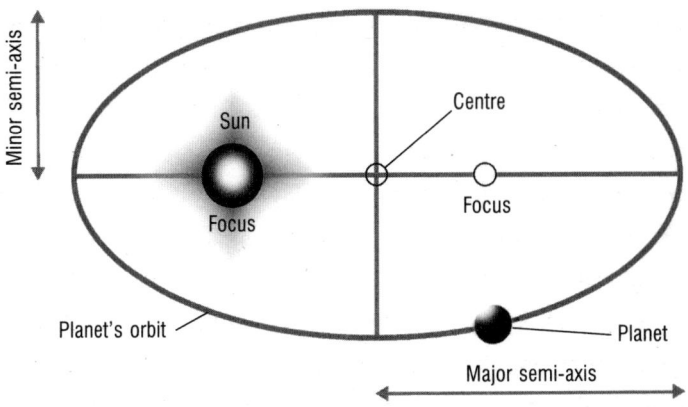

Kepler's first law asserted that the orbital path of a planet will always be an ellipse.

Time and motion studies

Kepler pursued his elliptical studies in the early decades of the 1600s, from which emerged his celebrated Three Laws of planetary motion, the first two of which were published in 1609.

The First Law regarded the shape of a planet's orbit, which, Kepler asserted, was an ellipse. What is more, the sun was not at the centre of the orbit, but at one of the foci. In the case of the earth, the orbit is actually quite close to a circular motion, something that can be judged using a number called the 'eccentricity' of the ellipse, which measures how 'stretched out' the shape is. In technical terms, it is measured as:

$$\sqrt{1 - \frac{b^2}{a^2}}$$

where a and b are the semi-axes. In the case of a circle, $a = b$ and the eccentricity comes out as 0, not far off the Earth's orbit which has an eccentricity of around 0.02. In fact, the distances involved are (for a) 149.60 million kilometres and (for b) 149.58

million kilometres. With a naked eye this ellipse looks very much like a circle.

In contrast, with an eccentricity of around 0.97, the orbit of Halley's Comet little resembles a circle, having a very long, thin appearance. Its semi-axes are very different from one another: (for a) around 2,663 million kilometres and (for b) around 623 million kilometres.

The upper limit on the possible eccentricity of any ellipse is 1, and the closer to 1 it gets, the more stretched out the ellipse becomes. Of the planets in our solar system, Mercury has the largest eccentricity, at around 0.21, inheriting the prize after Pluto (with an eccentricity of 0.25) was demoted from planetary status in 2006. Mars, which was the focus of Kepler's attention, has an eccentricity of around 0.09.

Kepler's First Law represented a huge breakthrough in our understanding of the geometry of the universe. But it was not the whole story. In particular, it did not address the *speed* of the orbiting planet. Brahe's astronomical observations convinced Kepler that planets did not travel at a constant speed. In fact, when they are nearer the sun they travel faster than when further away, behaviour addressed by Kepler in his Second Law. He imagined the planet joined to the sun with a long rod. Then, he said, the area swept out by that rod every minute is always the same, irrespective of what stage of its orbit the planet has reached. There is nothing special about the duration of one minute – any fixed length of time would work just as well.

Kepler's Third Law followed the first two a decade later, in 1619. It related the length of the planetary orbit to the time taken to travel it. As one would expect, a longer orbit takes more time

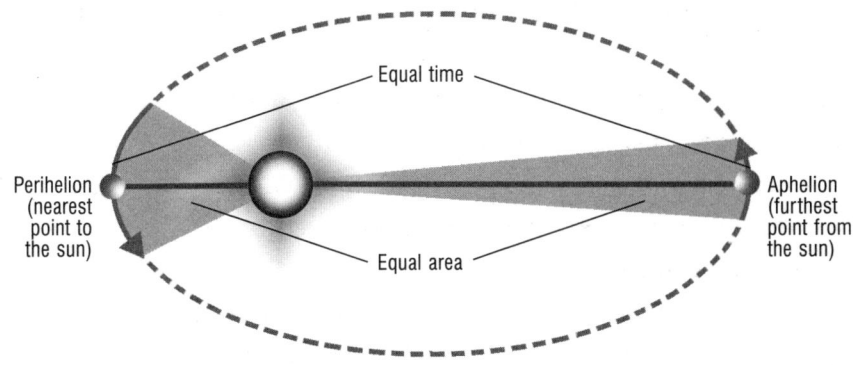

Kepler's second law guaranteed that the area swept out by a planet in a fixed length of time will be the same at every stage during its orbit.

to travel, simply on account of the extra distance involved. On the other hand, planets with shorter orbits travel faster than those on long ones. So the exact relationship is not immediately clear. Kepler established the answer with a subtle law. Taking the time of the planet's orbit as T – in other words, the length of its year – and the length of its semi-major axis as a – that is, the maximum distance the planet reaches from the centre of its orbit – then T^2 is proportional to a^3.

The values of T and a may vary from planet to planet, of course. But Kepler stated that there must be a special number, K, deriving from the mass of the sun, so that for every planet, it must be true that $T^2 = K \times a^3$. In our solar system, K is around 2.97×10^{-19}. In fact, this is only an approximate result. An exact statement would also take into account the mass of the planet. But with the mass of the Sun a hundred thousand times greater than even that of Jupiter – the solar system's heaviest planet – for almost all purposes the planet's mass can be ignored. But in binary star systems in which two stars or black holes are locked in orbit around each other, Kepler's third law must be slightly adapted to incorporate the mass of both entities.

In the case of the earth, Kepler's Third Law connects our major semi-axis of around 149,600,000,000 metres to the length of one earth year (31,536,000 seconds). On Mars meanwhile, the semi-axis is around 228×10^9 metres. So, using Kepler's formula we can deduce that the length of a Martian year is around $\sqrt{2.97 \times 10^{-19} \times (228 \times 10^9)^3}$ seconds, which comes out at around 687 earth days.

The Newton effect

With his Three Laws, Kepler had accurately captured the motions of the planets and made a profound contribution to astronomy. What he did not explain, however, was *why* they should travel along these elliptical orbits, at these particular speeds. The explanation for those phenomena had to await Isaac Newton (1642–1727) later in the century.

While Copernicus, Galileo and Kepler had all suffered from their clashes with authority, Newton was very much an establishment figure, his own unorthodox religious views notwithstanding. At various times in his life he was Lucasian Professor of Mathematics at the University of Cambridge, a Member of Parliament, Master of the Royal Mint and President of the Royal Society, for which services he was knighted by Queen Anne. He was also deeply interested in optics and telescopes (see *Let there be light*).

In 1687 Newton made a decisive contribution to our understanding of the solar system with his Law of Universal

Gravitation, which appeared in his *Philosophiae Naturalis Principia Mathematica* (*The Mathematical Principles of Natural Philosophy*, but generally referred to as just the *Principia*). The law states that all objects with mass are gravitationally attracted to one other. This will include planets, moons and stars, as well as the legendary Newtonian apple. But this force of attraction diminishes the further away the two objects are.

More specifically, the gravitational force between two objects is *proportional* to each of their masses – m and M – but lessens in proportion to the square of the distance between them, referred to as r^2. That is to say that the gravitational force is proportional to the product of the two masses and inversely proportional to the square of the distance separating them.

To work out the gravitational pull of m on M at distance r apart, one other ingredient is needed: the so-called universal gravitational constant G – a numerical statement of consistent proportionality, whose value is around 6.67×10^{-11}. Armed with this information, the gravitational pull on each mass is equal to:

$$\frac{G \times M \times m}{r^2}$$

In the case of the earth and the sun, the sun's mass (M) is around 1.99×10^{30} kilograms, while that of the earth (m) is around 5.97×10^{24} kilograms, and the average distance (r) between them is 1.49×10^{11} metres, giving a force of roughly 4×10^{22} newtons. (Units of force are named after Newton.) The numbers are astronomical, and not just in a literal sense. But as the US quantum physicist Richard Feynman commented, the gravitational principle that Newton formulated 'is simple, and therefore it is beautiful'.

Newton was further able to use mathematical techniques of his own discovery to derive Kepler's Three Laws from his own theory of gravity. With this step taken, the long journey from a geocentric perception of heavenly bodies rotating around the earth reached an impressive scientific understanding of orbits, distances and the relationships of sun, planets and moons. It was not everything of course; for one thing, Newton had not been able to account for what caused gravity. For another, there remained the interplay of three, rather than two, bodies in space, such as the sun, a planet and a moon (see *One small step* ...). Most importantly though, the earth had finally found its correct place in the solar system – a possibly humbling discovery, but one depending on some awe-inspiring observational science combined with impressive mathematical insight.

THINKING MACHINES

The mathematics of
machine learning

The concept of creating intelligence in a machine is a daunting and – if we believe Hollywood – even doom-laden prospect. Do we human beings really want assemblages of metal, plastic and electronics to be able to *think* as we do? It is no wonder that science-fiction writers and screenwriters have regularly visited the theme to pit hapless humans against malevolently smart machines. Usually the humans win out, owing to some lucky fatal flaw in the make-up of their adversaries' systems. But the whole genre testifies to an enduring unease, perhaps not helped by the bold claims made at the 1950s conference at Dartmouth College that founded artificial intelligence (AI) as a discipline: 'Every aspect of learning or any other feature of intelligence can in principle be so precisely described that a machine can be made to simulate it.'

That declaration seems extraordinarily ambitious; after all the human brain is an astonishingly complicated object, capable of tasks unmatchable by any other creature or system we know of. So AI researchers have an immense amount of ground to make up. On the other hand, it is perhaps surprising the degree to which artificial intelligence has already penetrated our lives in its half-century of existence, without – as yet – scaring us to death.

The truth is that sidelining the human race is not on anyone's wish list. While there are many threads of AI research, most are geared towards more immediately practical goals than laying the foundations for some future dystopia. One branch of the subject goes by the name of 'machine learning' and is the art of training computers to master 'human' skills such as recognizing people's faces or transcribing spoken language. From finding exactly the right mix of concrete for a particular construction task to the personalized online adverts thrown up by internet search engines, there is plenty of evidence that today's machines are already doing their daily homework and, yes, learning.

Machine learning

Human learning has a great deal to do with experience: a child recognizing her mother in a crowd of faces, an ornithologist identifying a birdsong, a perfumier naming a scent, even the seemingly simple tasks of identifying day-to-day objects such as chairs, cups, telephones. The wonder of the human brain is that we are able to intuit categories from experience: we can usually identify an object as a chair, cup, telephone with no extra effort, even when we have not seen that particular design before.

This type of learning is not something that comes naturally to computers; they are excellent at following long lists of detailed instructions – algorithms – but for many types of learning such an explicit procedure is not the right approach. If we wanted to build software capable of recognizing a human face, for instance, we might be tempted to encode instructions that translate as:

'This is what an eye looks like. There should be two of them, and the horizontal distance between them should be approximately *x*. This is what a nose looks like. It should be between the eyes, vertically offset by a distance of approximately *y* ...' This is not how children learn to recognize human faces, and for the most part it is not the approach that programmers use either because it is extremely likely to fail. These instructions would be flummoxed when faced with an unusual face, perhaps very young or very old, or someone with their eyes closed, or with a birthmark or an eye-patch, or simply a face photographed from an unusual angle or in imperfect lighting conditions – any of these attributes could throw the program off. Somehow, the machine needs to be enabled to learn for itself and then apply the results of that learning. One manifestation of that is the facial detection available on many of today's digital cameras – an example of technology catching up with humans.

There is another side of the AI story too, where we are not looking to machines to replicate our abilities but rather to apply themselves to tasks that we do not particularly excel in. Working out the precise mixture of ingredients for particular applications of concrete turns out to be an extraordinarily difficult task. Here, in the construction industry, machine learning is already taking the burden, leading to real advances in understanding that are highly practical and time-saving for engineers.

Yes or no? The beginnings of AI

One of the simplest and oldest instruments in machine learning is a perceptron – not a gadget, but a program conceived in 1957 by Frank Rosenblatt. It applies a binary principle, in that it divides objects into one of two categories. For example, in a photograph of the night sky, some of the lights depict nearby stars, while others are distant galaxies. It is not completely straightforward to tell one from the other, but there are patterns: stars tend to be rounder, while galaxies are typically longer and thinner. What makes this task difficult is that this is not an exact rule: some galaxies seem fairly circular, while gravitational lensing may distort the light coming from a star, to give it a flatter appearance.

To learn the difference, a machine cannot start from nowhere. Instead it needs a set of training data. This might comprise a collection of spots of light and three pieces of information about each of them: its width (i.e. the longest distance across it in the photograph), its height (the narrowest distance across it), and whether it is a star or a galaxy. The first two of these can be plotted on a graph, giving a collection of data points. The perceptron then tries to draw a straight line – a decision boundary – on the graph, with stars to one side and galaxies to the other. As new data points are introduced, the perceptron will apply the rule it has learned, according to which side of the line the point lies. If it slips up, the perceptron *learns* by slightly adjusting the steepness or position of the line.

The perceptron is not limited to two coordinates; if we wanted to incorporate a third reading for light-intensity into the stars and galaxies example, the perceptron would plot points in three-dimensional space and try to split the two categories with a plane. All the same, by modern standards the perceptron is a rather primitive program. To start with, it will only work when the decision boundary between the two categories is a nice straight line or a perfectly flat plane. In many practical situations this is simply not the case.

An alternative program might investigate neighbours to a new point: if out of five neighbours, four were galaxies and one a star, then the machine might 'guess' that the new point is a galaxy too, by means of a majority vote. Although conceptually simple, this in fact can be a very effective process, and it is untroubled by how wiggly the decision boundary between the two categories is. A major disadvantage, however, is that such a program is computationally highly intensive: to make a judgement about the new point requires searching through all the previous ones, of which there may be many millions, to find those closest to the new point.

To overcome these problems and tackle more sophisticated tasks than simple binary classification, today's theorists have built an array of more sophisticated approaches that can combine different types of learning within a single system. And the most famous of these – neural networks – are inspired by the architecture of the human brain.

Neural networks

The brain contains a network of neurons – specially adapted cells that receive and emit electrical signals. Neurons have input points known as dendrites, where they receive signals from neighbouring neurons. Some of these signals 'excite' the neuron (i.e. increase the voltage across it), while other 'inhibit' it (decrease the voltage). Critically, every neuron has a particular threshold, so when the voltage exceeds that level the neuron will fire and go on to excite (or inhibit) those other neurons to which its terminals (outputs) are attached.

The human brain encompasses 100 billion neurons in a complex network, many of them linked to tens of thousands of others. At crucial points, sensory neurons receive their inputs not from other neurons but from signals from the outside world via our senses of touch, taste, sight, smell and sound. Similarly, the outputs of specialized motor neurons are not other neurons, but muscles, which are prompted to flex or relax.

Despite the brain's immense complexity, today's biologists have a good idea of its micro-operations at the level of individual neurons. The greater mystery is in how large networks of neurons give rise to such phenomena as thoughts and emotions. One approach to this question is through the study of neural nets, artificial systems that mimic the structure of the brain, albeit with some variation. Essentially, they are collections of artificial neurons (also known as 'Threshold Logic Units') connected with channels.

So far, our discussion has been relatively mathematics-free – but we finally get to numbers in the context of examining how a single artificial neuron works. As with its biological counterpart, an artificial neuron has several input channels along with one output channel that feeds the inputs of one or more other neurons. Each input is a number. If there are three input channels, we can assemble their values into a column vector. So if our three inputs are 1,1, and 0, the input vector would be $\begin{pmatrix} 1 \\ 1 \\ 0 \end{pmatrix}$.

The artificial neuron does not simply accept these inputs

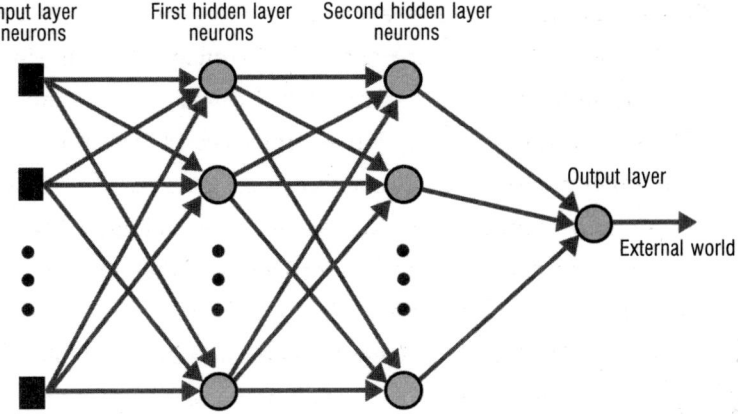

Input layer neurons

First hidden layer neurons

Second hidden layer neurons

Output layer

External world

A schematic diagram of an artificial neural network.

wholesale, for at each input channel is a number called a weight. If we imagine these as 2.5, -1 and 0.5, the weight vector would

be: $\begin{pmatrix} 2.5 \\ -1 \\ 0.5 \end{pmatrix}$

The exact values of the weights will vary from neuron to neuron (and they may change from time to time too).

From the inputs and the weights, a single number known as the net input is produced, calculated by multiplying the value of each input by the weight at that channel, and then adding the results. This is a familiar operation from algebra, where it is known as taking the 'dot product' of the two vectors. Using our figures above would generate:

$$\begin{pmatrix} 2.5 \\ -1 \\ 0.5 \end{pmatrix} \cdot \begin{pmatrix} 1 \\ 1 \\ 0 \end{pmatrix} = 2.5 \times 1 - 1 \times 1 + 0.5 \times 0 = 2.5 - 1 = 1.5$$

This net input determines what the neuron fires as its output, but there are various methods – or more properly activation functions – that it might use. The simplest procedure is to output 1 if the net input lies within some predefined range, and 0 otherwise. It might be that our neuron outputs 1 if the overall input is positive, and 0 if it is zero or negative, so with our example, with a net input of 1.5, the neuron would output 1.

Neither an artificial neuron nor a biological one is inherently complex. But complexity attaches to them en masse when they are assembled into a network. A typical neural net has artificial neurons of three types. Those in the input layer receive their inputs from the external world, perhaps via the keys of a keyboard, the pixels of a digital camera, or the frequency channels of a microphone. At the output layer, the outputs feed into the external world, perhaps via a computer screen, a speaker or a robotic arm (see *Automata and articulation*). Between these two may lie several layers of hidden neurons.

These artificial neural networks have the capacity to learn, to be trained. For instance, we might try to train one to recognize handwritten letters drawn on a grid, each pixel of which is wired to one input neuron. In a standard neural network, information only flows forwards through the network – from the sensors to the input layer of neurons, then cascading through the hidden layers of the net with pieces of data combining and being weighted in various combinations, before (in this case) a single output neuron finally emits the result of the computation. As we have seen, neural networks convey information as numbers. So, for our letter-recognizing network, we will need to employ the standard alphanumeric code: $1 \to A$, $2 \to B$,

$3 \to C$, etc. It may be that the network produces an overall output of 14, meaning that it is identifying the written letter as an N.

The 'target output' is the term for the correct answer. In this case, we may have been hoping for an M, so the net has made a mistake. Now we arrive at the crucial moment where, we hope, the net can learn from its mistakes. The procedure involves amending the weights at each neuron so that re-running the computation on the same input will now produce the correct output. But the details here are important. After all, one could simply rig the system to produce 'N' every time. Needless to say, this is no kind of learning; if the net had already mastered 'L', all that work will be undone. So the weights are moved to a *nearby configuration* that will output the correct answer.

In an ideal scenario, the net would correctly assign an output to each possible configuration of input pixels. The complete raft of matches between all possible inputs and their correct interpretations is called the 'target function', and it exactly captures the behaviour that we want the net to master; or more realistically, we want to *minimize the distance* between this target and the net's actual output.

Luckily there is a deep well of mathematics to tap into here, for analysing and minimizing the distance between two functions is a topic mathematicians have studied for several centuries. We can measure the net's overall rightness or wrongness with an error function, which we can call E. On a particular run, if the desired target is the number 7, and what the neuron actually outputs is 5, then we have a clear way of measuring the extent of the error: $7 - 5 = 2$. As a general principle, if the target is s and actual output is t, then the error is $s - t$.

Now, this value is only a measure of how well the net performs one particular task (recognizing one particular particular hand-written letter). We can get other values from all the other tasks the net might try, with greater or lesser success. It would be desirable to add up all these errors to gauge the neuron's overall effectiveness. However, there is a minor obstacle here: if the error on another task is $4 - 6 = -2$, then that error in combination with our above-mentioned error will cancel one another out. For this and other reasons, it is preferable to use the error formula squared $(s - t)^2$, which would be 4 in each of the above cases. We can add up these values across all the net's tasks, to get an expression for the total error of the net as it currently stands, a single number, perhaps $E = 128$.

Once we have arrived at a value for E, the question is: what to do with it? Clearly, improving the net's performance means bringing the total error E down. The way to do this is to adjust the neurons' weights, and it is enlightening to adopt a geometrical approach here.

To illustrate the idea, we might focus on a single neuron, with two input channels. We can imagine its two weights as representing the coordinates of a point on a map. The error function E can be thought of as the height of the terrain at that point. Altering the weights, individually or together, can bring us uphill or downhill as the neuron gets better at its job, or worse. A critical factor is how closely E depends on a given weight. It may be highly sensitive to changes at one channel, while the other may make only a small amount of difference. In geometrical terms, we may think of this as the steepness of the landscape, which might rise steeply to the north but only gently to the west.

With this geometry in mind, we can introduce a geometrical procedure called the 'delta rule', which aims to adjust the weights in the direction of steepest descent in the value of E. In other words, the delta rule will alter the weights so as to produce the greatest improvement in accuracy, for the smallest adjustment.

To train a neural net, theorists of artificial intelligence proceed backwards through the neurons, starting with the output neurons, tweaking their weights according to the delta rule, to produce the correct overall answer. This procedure is known as 'back-propagation', and it is the standard way in which artificial neural networks learn.

The results of the theorists' efforts are, in many cases, already embedded in applications and processes today, such as speech-recognition software. Occasionally artificial intelligence also grabs the headlines with one of its more eye-catching experiments, as with recent inventions of driverless cars independently sensing their way using neural networks or 'artificial noses' that can detect scents and flavours. The whiff of science fiction may still be present in some of the more exotic endeavours – but there is plentiful evidence that artificial intelligence is already changing our world. As to where that change may end up – who can say?

UNTANGLING THE STUFF OF LIFE

The mathematics of DNA

In 1953 a revolution in our understanding of life was ushered in by the duo of James Watson and Francis Crick – although you would have been hard pressed to guess it from their original paper, with its gentle suggestion that 'this structure has novel features which are of considerable biological interest'. In this work, they gave modern science perhaps its most alluring shape: the structure of 'deoxyribonucleic acid' – DNA – as a double helix. In doing so, they affirmed that DNA was indeed the stuff of life, the molecules within the nuclei of our cells that govern heredity and encode our evolutionary legacy. This breakthrough laid the groundwork for the future science of genetics, and its potential to screen for and repair genetic disorders.

Since then, genetic science has moved in leaps and bounds, reaching new sophistication with the emergence of genomics, the study of the genetic make-up of whole organisms. In 1977 the genome of a bacterium was decoded; less than 25 years later, the first draft of a human genome had appeared, described by US President Bill Clinton as 'the most important, most wondrous map ever produced by humankind'.

The next step – to read and understand the DNA sequences that researchers have extracted – requires mathematical and computational ingenuity, but it holds the promise of great scientific and medical rewards.

The geometry of the double helix

Centuries before biochemists shot the helix to fame, its form had intrigued mathematicians. A helix (plural *helices*) is a curve, but more particularly it is a space-curve – its home is in three dimensions rather than on a flat piece of paper. Nevertheless, a helix is built from the two most fundamental shapes in mathematics: a circle and a straight line. To create a helix, we can imagine a circular metal hoop along which a bead is travelling at a fixed rate; if we now move that hoop forwards in straight line, at a consistent speed, the course traced by the bead creates a helix. A classic spiral staircase is, in fact, a misnomer, being a helix rather than a spiral.

More precisely, the geometry of the helix is determined by three factors. The first is the size of the circle. In a molecule of DNA, its radius is about 1 nanometre (nm) or 0.000001 millimetres. The second consideration is the space between successive twists of the curve. We can imagine this being decided by the speed at which our circular hoop travels: quickly and the curves will be few and far between; slowly and the helix will be tightly coiled. In DNA, successive curves, or branches, are around 3.3 nanometres apart. The final consideration is whether the bead travels around the track clockwise or anticlockwise (viewing from behind as the hoop moves away away). Clockwise travel generates right-handed helices, while anticlockwise movement

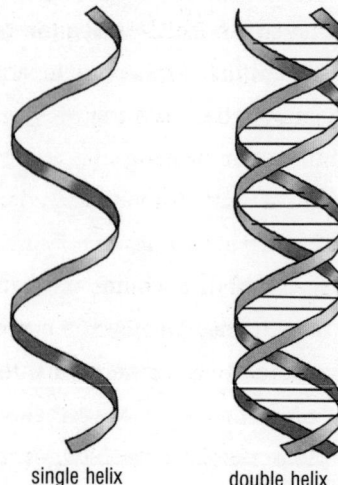

A molecule of DNA takes the form of two intertwined helices, locked together by chemical bonds.

single helix   double helix

produces a left-handed helix. DNA of either orientation is possible, though right-handed helices predominate in nature.

The geometry is useful in understanding a major practical problem that DNA poses. Each human cell contains around 2 metres of DNA, most of which appears within the nucleus. This has been likened to 200 kilometres of fishing line being stuffed into a basketball. In such a scenario, the risk of the whole thing deteriorating into a hopeless tangle appears dangerously high. To guard against this, the nucleus contains an army of enzymes whose job it is to look after the DNA, copying it, building proteins from it, while also preventing it from becoming knotted or torn. One protective mechanism they use involves the different ways the DNA molecule can twist and turn.

A molecule of DNA is not just a single helix, of course, but rather two locked together. To model this, it is useful to move beyond a simple curve, and instead regard the molecule as a flat ribbon, with two edges running either side of a central axis.

There are two essentially different ways in which such a ribbon can bend. The first move is the 'twist', which happens when the two edges swap sides across the central axis – as would happen when pulling a real ribbon taut and then twisting one end by 180°. The important observation here is that although the edges change sides, the central axis itself continues to run in a straight line. The helical nature of DNA means that it comes naturally twisted; however, enzymes have a tendency to introduce new twists as they work on it, and too much twisting can place the molecule under strain and cause damage.

Luckily, there is a safer way in which a ribbon can turn, which goes by the name of 'supercoiling'. An everyday example is the telephone cord. Technically, supercoiling takes place when the ribbon's central axis crosses over itself. Even with the ends of the ribbon fixed in place, a twist may convert into a supercoil, or vice versa: this is how supercoiling protects the DNA from over-twisting.

Mathematically, this process can succumb to a neat algebraic description. If T is the total number of twists – with each anticlockwise twist contributing $+1$ to the total and each clockwise twist -1, and S is the number of supercoils (counted in a similar way), then neither S nor T is a fixed quantity. If you give the ribbon a shake, one number may go up and the other down. But the total $S + T$ will remain the same, so long as the ribbon's ends are fixed in place. This total is known as its 'linking number'.

From curves to ladders, from genes to genomes

Simple models such as smooth curves and ribbons can be useful for understanding the geometry of DNA. But in reality, the double helix is a much blockier object, built from special molecules called bases. There are four types, which come in two pairs: cytosine (C) whose partner is guanine (G), and adenine (A) with its opposite number thymine (T).

The structure of the DNA molecule is a string of these bases: ...GAGCT... and in humans this string is over a hundred million bases long. However, it is not alone; it has a companion, the second strand of the double helix. Where one has a C-base, the other has G, and where one has A, the other has T. So, for the sequence above, the second string would read ...CTCGA.... It is these base pairs that create the impression of the rungs of a twisting ladder.

The sequence of base pairs encodes information about the organism, but they form the very lowest level of the informational structure. A higher unit is the gene, a section of DNA that functions as a self-contained reproductive unit, passed on from parent to offspring, and treated as a single entity within the cell. Each gene comprises somewhere between a few hundred and several thousand base pairs, and the overwhelming majority of genes encode the structures of the numerous proteins used in different parts of the body. Proteins are the workhorses of the body, holding every cell together; they are the scaffolding that allows cells to keep their shape rather than collapsing. The armies of enzymes that run the cells, breaking down foodstuffs into sugar, untangling the DNA, and running the instructions encoded within it, are individual molecules of protein too.

A chromosome is a complete length of double helix DNA with its component genes. Human cells have 46 of these, which come in pairs (one from each parent). However, genes themselves only constitute around 2 per cent of our chromosomes, with the remaining non-coding segments commonly known as 'junk DNA'. Whether or not this is a fair description remains a topic of debate.

The year 2003 saw a milestone in the history of genetic research, when the US-led Human Genome Project completed its primary goal in mapping a complete sequence of the 3 billion base pairs of a human genome – in other words, compiling the complete catalogue of all the genes on all its chromosomes. This was certainly a magnificent achievement. But the discovery that humans have only around 23,000 pairs of genes caused a shock. It seemed too small a number of parameters to describe something as complex as a human being. But the mathematics tells a different story, because if we use this number to produce a crude estimate for the possible number of genetic humans, the answer comes out at 2^{23000} – that's 23,000 twos multiplied together. This assumes that each of the 23,000 human genes will adopt one of only two forms, known as its two 'alleles'. Essentially, this gives us a sequence of 23,000 switches, each of which can be in one of two positions.

In fact, the figure of 2^{23000} should not be taken too seriously; the true number is likely to be higher, since recent research suggests that even identical twins have slight genetic differences. All the same, at around 10^{6900}, this number enormously exceeds the estimated number of atoms in the known universe (approximately 10^{80}). This is a powerful illustration of an important principle: that combining a seemingly small number of variables in different ways can give rise to a staggeringly complexity.

How to read the book of life

For the Human Genome Project to be truly useful, scientists have had to learn to read and analyse the genome, a project in progress. For one thing, despite frequent media reports of the discovery of a 'gene for X', human characteristics such as nose-length, or the vividness of visual memories, are seldom governed by single genes, but by complex combinations of them. It is only by reading the results carefully that we have been able to begin to search for the genetic basis of hereditary diseases, or make judgements about how closely our genes resemble those of our cousins across the animal kingdom, and beyond. Gene sequencing is surely one of the most exciting technologies to develop in recent years – and in order to work, it needs certain mathematical puzzles to be solved. (See also *Syphilis and Christmas lights.*)

The issues can be best explained by way of examples, so let's suppose a scientist has sequenced a gene from a fruit-fly and wants to know whether humans share this gene. What is needed is a way to search the human genome for any occurrence of the fly's sequence. If that were all there was to it, the job would be straightforward. The complicating factor is that DNA changes over evolutionary time, so base pairs can end up inserted, deleted or exchanged. So we cannot expect *precisely* the same sequence to appear. What is needed is a way to search for a *similar* stretch of human DNA. This is a much tougher task than finding a perfect match.

The primary tool that today's biologists use for this is called BLAST – the *Basic Local Alignment Search Tool*, developed by a team of US researchers in 1990. To illustrate how it works, let's suppose the fly's gene begins AGCGTC ... and we want to compare this to a promising stretch of human DNA, which reads ACCTGTC ....

A critical step to allow for the possibility of deletions or insertions is to introduce a new symbol to the alphabet of bases, standing for a space: _. With this, we can explore different alignments between the two sequences. One possibility might be:

| Fruit-fly gene | A | _ | G | C | G | _ | T | C | ... |
| Candidate human DNA | A | C | _ | C | T | G | T | C | ... |

Now, we want some way to assess the alignment, in a way that quantifies the degree of similarity. The standard way to do this is to apply a method of scoring, in which each matching base adds +5 to the score, while each mismatch is penalized by −4, and the net result is totalled. (These are the default settings; more complex scoring systems are also possible.) Under these rules, our example alignment would yield a final score of +4.

Of course, there are a great many possible alignments, depending on where the human sequence is started and where spaces are inserted. BLAST's aim is to find the alignment with the highest possible score. In our selection, for example, deleting the first gap in both sequences would increase the score by +4 to +8.

Unfortunately, a complete search of every possible alignment is well nigh impossible, for the numbers grow too quickly. Thus some ingenuity is required to filter the results, and the means to do this is by comparing short sections first and rejecting all alignments that fail to reach some threshold score. (In deciding this, it is important to take into account the probability of matches arising through chance alone.) Then the remaining candidate alignments are extended, in multiple ways, and compared to the target again, until – with luck – just one candidate remains.

This process, despite the ostensibly laborious computations it seems to imply, underlies some of the most exciting science today. The individual genes responsible for debilitating conditions such as cystic fibrosis and the hereditary neurological illness Huntingdon's disease have now been identified, which already allows risk-assessment and screening, and may in time lead to gene therapy in which defective genes can be mended by the insertion of new DNA.

Meanwhile, we continue to learn more about our place in the world, where we came from, and the development of species. It was perhaps inevitable that, with the human genome mapped, attention would turn to our nearest relative in the animal world, the chimpanzee. The ongoing Chimpanzee Genome Project in the United States, starring a male chimpanzee named Clint, suggests that 96 per cent (at least) of our genetic material is almost identical to that of our closest cousins. The numbers speak loudly of a shared ancestry, millions of years ago, a finding Charles Darwin would surely smile upon.

BALLOT BOX PARADOX

The mathematics
of elections

Dictatorships and tyrannies do, in some respects, make for simplicity. There is little need to pay any more than lip service to the intricacies of representative government, and consequently there is no need to explore the pros and cons, and the relative fairness, of competing electoral systems. Nevertheless, tyranny is not much of a temptation, and most of us for good reason – and for all its faults – hang onto some version of democratic government as the means to choose our leaders and representatives.

In a functioning democracy, voters register their preferences for candidates at election time and the job of the electoral system is to amalgamate all these opinions into a final outcome. Around the world, and throughout history, numerous different systems have been used, because this central task is infinitely harder than it first sounds. To understand the difficulties, we can appeal to the branch of mathematics known as social-choice theory. Working through some examples of voting systems, we can see some of the surprising and counterintuitive situations that sometimes emerge from the ballot box.

Getting past the post

To begin with, let's suppose an election has five candidates – Albert, Beatrice, Cleo, Dana and Edward (A, B, C, D and E). My own order of preference might run BEDCA, but my wife might prefer DBACE, and others will have different preferences. The simplest way to find the overall winner in such a system is to ignore everything except voters' first preferences (in my case B, and my wife's D). Then we accept the candidate who has the most votes overall. This system is called 'plurality voting', or 'winner-takes-all'. In UK parliamentary elections, it is known, somewhat inaccurately, as 'first past the post'. The trouble is that there is no 'post' to get past as such. In principle, a candidate may win a plurality election with a small proportion of the votes, so long as the other candidates' tallies are even smaller. In the imagined result of our example, candidate A wins with only a quarter of the votes:

A	25%
B	21%
C	20%
D	19%
E	15%

The effect becomes more pronounced the more candidates there are. For the sake of argument, imagine an (admittedly unlikely) election with 1,000 candidates, who each receive one vote, except for candidate A who receives two votes and candidate Z who receives no votes. Under the plurality system, candidate A wins the election, despite having only 0.2 per cent of the vote (and possibly being the least favourite candidate of the remaining 99.8 per cent of the population).

Running off

The alternatives to plurality voting work by taking into account voters' preferences beyond their favourite candidate. The best known of these is Instant Run-Off Voting (IRO), also known as the Alternative Vote. The idea is that a winner will only be declared once they have above 50 per cent of the vote – thus overcoming the main problem with plurality. This could be accomplished with several rounds of voting. Imagine a three-candidate election. After one round of voting, the results stand at:

A	38%
B	47%
C	15%

Since no candidate has 50 per cent, no-one has yet won. Instead, the lowest-scoring candidate, C, is eliminated. Now, we have a second round of voting, with only two candidates. Let's make the assumption that all the people who voted for A or B in the first round will stick with the same choice this time around. So the final outcome depends on how the 15 per cent of C-voters now split between A and B. If, say, 5 per cent now opt for A and 10 per cent for B, then the overall winner will be B with 57 per cent of the vote to A's 43 per cent. (Elections with more candidates may require further rounds.)

This example is run-off voting, and the only difference with the instant run-off system is that voters register their second (and higher) preferences at the initial election, to save everyone trekking back to the ballot-boxes again.

Condorcet winners and Arrow's Paradox

Instead of focusing on specific voting systems, mathematicians prefer to think about the issues more abstractly. For a start: what are the conditions that we might want a voting system to satisfy? Important in this respect are the Condorcet criteria, so named after Marie de Caritat, Marquis of Condorcet, who investigated them in 1785.

An electoral candidate satisfies the condition of being a Condorcet winner if he/she would beat every other candidate in a two-way election.

This certainly sounds like a good qualification for any winner, but, as always in social choice theory, things are not as straightforward as they seem. Imagine a three-candidate election (A, B, C) where voters rank all candidates in preference order, producing a percentage breakdown of:

CAB	49%
BAC	41%
ABC	10%

(We're assuming, for this example, that no-one votes BCA or any other permutation.) In this election, the Condorcet winner is A. To see the result of a two-way election against B, simply delete the Cs from the above table and then add up the results of the two rows for AB against the one for BA. This shows that A would beat B by 59 per cent to 41 per cent. Similarly, A would defeat C by 51 per cent to 49 per cent. It is by no means obvious, however, that A is the natural winner of this election. In particular, candidate A received by far the fewest first-preference votes, and so would come last in a plurality vote – and be eliminated first in an IRO election!

However, Condorcet criteria present an additional problem, in that most elections do not actually generate a Condorcet winner. For instance consider this result in preferential voting:

ABC	40%
BCA	35%
CAB	25%

In the Condorcet two-way run-offs, A beats B by 65 per cent to 35 per cent, while B defeats C by 75 per cent to 25 per cent; but C beats A by 60 per cent to 40 per cent, meaning none of the three candidates is a Condorcet winner. This cyclical arrangement is known as Condorcet's Paradox, and it means that, on its own, the concept of the Condorcet winner is insufficient for a viable electoral system.

Nevertheless, several Condorcet voting systems are known that are guaranteed to select the Condorcet winner where one exists. The most popular was invented by Markus Schulze in 1997, and although no country uses Condorcet voting for its governmental elections, the Schulze system has been adopted by various private organizations, including Wikimedia, Ubuntu, and even MTV for deciding which videos to show.

By contrast, a Condorcet loser is a candidate who would lose every two-way run-off – though, as the previous example shows, there need not be a Condorcet loser. More surprising is that, even where one exists, that candidate may end up winning the election. Let's imagine another possible breakdown of preferential voting:

ABC	45%
BCA	30%
CBA	25%

(Again we're assuming that other arrangements, such as ACB, receive no votes.) Here, A is the Condorcet loser, because in a two-way run-off, both B and C beat A by 55 per cent to 45 per cent. Yet, in a plurality vote, A would win.

A Condorcet loser cannot, however, win an IRO election, since the final round will be a two-way match-up, which the Condorcet loser is, by definition, guaranteed to lose. In our example, C would be eliminated in the first round, and then B would win the second round against A, by 55 per cent to 45 per cent.

Besides plurality voting, IRO voting and its variants, and an assortment of Condorcet systems, there are other systems too – indeed, almost a confusion of potential choices. Why should it be so hard to arrive at a universal system that best represents the views of an electorate?

This is a question which anthropologists, historians, political scientists, and sociologists have all pondered. But mathematics offers striking insights too. One answer to this conundrum was provided by Kenneth Arrow in the mid-20th century. He first listed two common-sense criteria that any fair ranking system should satisfy:

• Unanimity. If every voter prefers candidate A to B, then A should be ranked higher than B overall.

• Independence of irrelevant alternatives. Whether or not A is ranked higher than B overall should depend only on voters' relative preferences of A and B, not on how they relate to some third candidate, C.

This train of thought, however, when pursued logically leads to a rather shocking conclusion: that the only system that satisfies both criteria is one where a single privileged voter determines the entire outcome: a fig-leaf election in a dictatorial system. If this individual ranked three candidates in the order ACB, then this would be the election's outcome, with everyone else's views ignored entirely.

Arrow's Paradox, as it is known, is the most famous fact in social-choice theory. It is also illustrative of mathematics' role in this arena: it cannot determine the 'best' voting system; that is a matter for people to decide democratically. What mathematics can do with great precision, however, is identify the trade-offs and difficult choices that need to be made.

Party lines and constituency boundaries

So far, our discussion has revolved around individual candidates in a quasi-presidential way. But in most political systems, there is an additional layer of democracy: political parties. If candidates' individual merits and demerits are accorded a lower significance than their party affiliation – which is very often the case in the hurly burly of politics – a system of proportional representation (PR) becomes possible for elected chambers.

At its simplest – and let's imagine three parties, the Advanced Party, the Backwards Party and the Central Party (or A, B and C) – the number of representatives each party is awarded can be directly proportional to its overall share of first-preference votes. There is certainly no mathematical difficulty. Any problems with it are political rather than mathematical. To start with, it requires a party-only system, eliminating the possibility of independent candidates standing. What is more, the result reflects national voting statistics only, severing the link between local representatives and geographical constituencies. Inevitably, therefore, various hybrid systems attempt to remedy the obvious deficiencies, for example by running elections locally using IRO voting, and then 'topping up' nationally, using PR.

PR also exhibits a tendency to produce no overall majority for any one party, meaning that governments are often coalitions. Again, this outcome is mathematically unobjectionable, but – especially in plurality systems – has been viewed by some with suspicion as leading to weak government.

Drawing the line

The practice of most modern democracies is to first elect representatives for individual constituencies, after which the party that has garnered the most representatives forms the government – or wins the right to negotiate with potential coalition partners should there be no overall majority. But this means that where the boundaries are drawn between constituencies becomes vitally significant.

Suppose a population of 1 million people is to be divided into two constituencies, each of which will elect a representative from either the Advanced Party (A) or the Backwards Party (B). To make the model easier, we'll suppose that people reliably vote in blocks of 100,000, making 10 blocks, each voting either for either A or B. In this particular population, A is only supported by two blocks, who all live in the far west of the country, while B has the remaining eight blocks:

A	A	B	B	B	B	B	B	B	B

If the population is divided up fairly into two constituencies, East and West, each of five blocks, the result will be the election of two B candidates, reflecting B's overall majority of votes:

A	A	B	B	B	B	B	B	B	B

But if the division were unequal, the result might be different:

A	A	B	B	B	B	B	B	B	B

Here, the west of the territory forms a constituency of three blocks and would elect an A-Party candidate, while the remainder forms a constituency of seven blocks, returning a B-Party candidate. In this scenario, the Easterners could fairly complain that their votes carry less weight than those in the west.

Clearly, in deciding constituency boundaries there is the potential, based on likely voting patterns of social groups, to skew the outcome of an election. For a constituency system to weight votes equally, each constituency should contain the same number of voters (or nearly so). But that is not all, for it remains possible to gerrymander the boundaries to favour one party or the other. Imagine a population of 15 blocks and 3 constituencies, in which most blocks vote for the Advanced Party:

A	B	B	A	A
A	A	B	B	A
A	A	A	B	B

Fairness demands equally sized constituencies of five blocks, but there are various divisions possible. With the following horizontal boundaries, each constituency will vote for A by a proportion of 3:2, meaning three A-Party candidates will be elected overall, with no representative for B. Supporters of B might well see this as unfair in giving them no voice at all:

A	B	B	A	A
A	A	B	B	A
A	A	A	B	B

But a redrawing of the boundaries into the following pattern would drastically alter the scenario. Now, the central district will elect a B-Party candidate by 3:2, while the eastern and western constituencies elect A-Party candidates (by margins of 4:1 and 3:2 respectively). This has the virtue of reflecting the two parties' overall support (nine blocks for A and six for B):

A	B	B	A	A
A	A	B	B	A
A	A	A	B	B

However, the Backwards Party's fortunes could be increased still further by some artful gerrymandering, as follows:

A	B	B	A	A
A	A	B	B	A
A	A	A	B	B

In this version, A-Party voters are packed into the eastern district (where they elect an A representative unanimously, by a margin of 5:0). But this allows B-Party candidates to win victories in the other two districts (by margins of 3:2 in each case), and steal an overall electoral victory!

Some countries, such as the UK, try to guard against gerrymandering by leaving boundary-drawing to independent bodies. The United States, unfortunately, has a long history of gerrymandering by both major parties. One upshot of this is the existence of some very strangely shaped constituencies, the most notorious being the Illinois Fourth Congressional District, shaped like a pair of earmuffs.

Mathematics or politics?

Elections, at whatever level, are inherently mathematical. After all, they revolve around counting and comparing votes. It is not surprising then that psephologists and political activists obsess about numbers, entering into agreement with minor parties to secure more second-preference votes, or poring over the data from exit polls. But the sheer mathematics of vote counting must always squeeze into a bigger political picture provided by local conditions and historical legacy. The negotiation is never cut and dried, and the nature of any representative system is summed up in Winston Churchill's aphorism: 'Democracy is the worst form of government, except for all those other forms that have been tried from time to time.'

Mathematically, the UK's first-past-the-post system could be extraordinarily unfair. If, say, two parties fought a general election and candidates for the winning party gained exactly 50.5 per cent of the vote in every constituency, the House of Commons would be inhabited entirely by the winning side, with no Members of the Opposition to reflect 49.5 per cent of the electorate. It never happens of course, because constituencies and voting patterns are not like that – though critics of the system would still point to the 'arbitrary' relationship between percentage of the total vote and numbers of MPs won. Yet recently, in a referendum, there appeared little public enthusiasm to change the system.

In the US Senate, the country's upper house, every state – no matter how small or how large – delivers two senators, a historic legacy that gives vast but under-populated Montana (with a population of less than 1 million) an equivalence with the mighty global economic powerhouse that is California (with its 37 million inhabitants). Purely mathematically, and in terms of proportionality, there can be no justification for this state of affairs. As with our first example of unevenly divided east/west constituencies, critics say that it delivers undeserved influence to the voters of 'heartland' rural states. Yet the system endures, because it is tied up with the history of respect for states' rights and the US Constitution, along with a powerful localism and distrust of 'big government'.

With social-choice theory, mathematics may provide democratic states with valuable tools of analysis, along with ammunition for critics and defenders of particular electoral systems. It may even yield the discovery of new systems altogether, such as the Schulze–Condorcet method. But the big political battles are usually won by other means.

OUR CGI WORLD

Triangulation and
computer-generated imagery

Mathematicians enjoy working with the formal, classic shapes such as the square and the sphere, which admit clean descriptions. But the world is full of more awkward shapes, and there is no simple formula for the outline of a hand or the contour of a human nose, for example. So when such a shape needs to be created – say for a computer game or animated film – programmers have no choice but to turn to approximate methods of modelling. Animators have been grappling with the issues since 1972, when Ed Catmull and Fred Parke produced a three-dimensional animated film of Catmull's hand, the first example of three-dimensional computer graphics.

In this task, though, another classic shape comes to the rescue. Although crafted into a high art, the underlying geometrical ideas of computer-generated imagery are uncomplex and elegant, and the commonest approach relies on breaking a surface down into the simplest geometrical shapes of all: triangles. And here there is an honourable history, for the mathematics of triangulation long predates these high-tech applications, having made its presence felt in many other contexts and applications, including human geography and analysing the spread of disease. Triangulations have been employed by mathematicians for centuries, since they encapsulate a great deal of geometrical information about surfaces, including the number and type of holes they contain (the particular interest of the branch of geometry known as topology; see *The hole story*), and are now a standard tool in computer modelling.

The method of triangulation proceeds along the following lines. Suppose we want to produce a computer model of a hand, to feature in a CGI (computer generated imagery) sequence of a movie. The first thing to do is pick any collection of points on the hand's surface. The next step is to join as many of these together as we can, with straight lines that *do not* cross over one another (or cut through the body of the hand). The result is a triangulation of the hand's surface. When a complicated curved surface such as this needs to be represented, the approach usually taken is to study the geometry of the original in enough detail to produce an accurate triangulation. Then the original shape can be forgotten, and the triangulation itself becomes the scaffolding for building the virtual representation. This final step involves filling in the skeleton with flat triangular panels. Of course, the immediate result may have a somewhat unnatural, angular appearance; this can be counteracted with clever artistry, by smoothing and shading, but also by making the triangular mesh fine enough.

Drawing the shape

There are innumerable ways to triangulate any given surface. This is clear from the simplest surface of all: the flat plane. It can be divided into countless combinations of large and small, fat

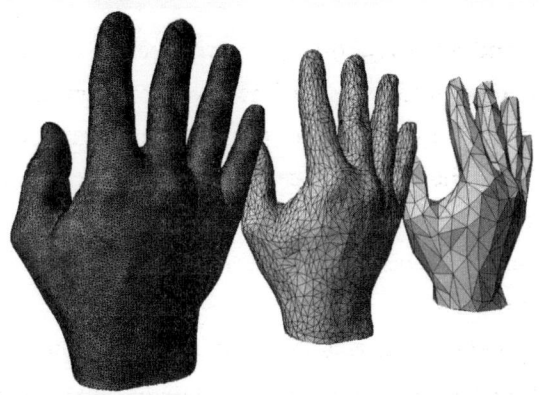

The finer the triangular mesh, the more closely the CGI hand resembles the real thing

and thin triangles. But some triangulations are more useful for computer modelling than others. The first obvious question is how fine the mesh of triangles should be. This is largely controlled by the density of the points, and the right answer will depend on the application: finer meshes produce more realistic approximations to the surface, but they also require more computational resources to manipulate.

However, the number of points is not the only factor. Once the points are specified, the next question is how to join them. Too many long, thin triangles will give the surface too much texture in one direction and not enough in another. To avoid this eventuality, the optimal approach is to use a special type of triangulation invented by Boris Delaunay in 1934. The definition of a Delaunay triangulation is that a circle drawn to connect the three corners of any one triangle should not contain any mesh-points except the three at its corners. Adhering to this requirement ensures that the triangles will not be too thin and will form a nicely spaced mesh.

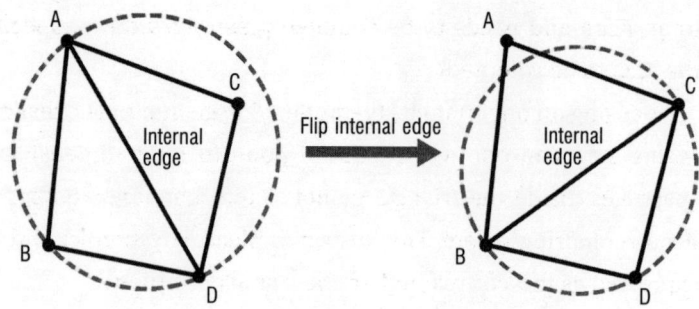

The left-hand triangulation is not Delaunay, since the circle on ABD contains the point C. Flipping the internal edge produces a Delaunay triangulation.

But how does one set about specifying a Delaunay triangulation? There is a neat trick which involves 'flipping' the triangles' edges. Begin with the points, and then draw in edges (lines) to form any triangulation at all. The chances are, of course, that it will not meet Delaunay's criterion. This will be evidenced somewhere by a triangle whose accompanying circle contains a fourth point, besides its three corners. These four points will consist of two triangles, and there is a simple way to improve the arrangement, by *flipping* the internal edge. Any triangulation, after enough flips, will conform to the Delaunay requirements, and programmers have turned this flipping process into quick methods for manufacturing Delaunay triangulations.

Filling it in

Let's return to our envisaged model of the hand. With the points of a triangulation chosen and the edges of a Delaunay triangulation specified, we have arrived at a scaffolding for our three-dimensional structure. The final step is to fill in the gaps with flat triangular panels. If we suppose our CGI hand belongs to an alien and needs to be coloured green, how do we specify the area to be coloured?

Focusing on one triangle, the underlying geometrical question is this: how can one tell whether a point in three-dimensional space lies inside the triangle or not – and therefore whether it needs colouring green? The answer is pleasingly simple, and is expressed as the 'convex hull' of the triangle's corners.

The mathematics of the idea is most easily conveyed using just two points, call them x and y Any points between them just form a line segment. The position exactly half-way between them is:

$$\tfrac{1}{2}x + \tfrac{1}{2}y$$

The point one third of the way from x to y lies at:

$$\tfrac{2}{3}x + \tfrac{1}{3}y$$

while the point three quarters of the way along is at:

$$\tfrac{1}{4}x + \tfrac{3}{4}y$$

The overall pattern is emerging here. In general, the points on the line segment between x and y are all those of the form $ax + by$, where a and b are positive numbers that add up to 1. It is this segment of line that is the convex hull of the two original points.

The same principle works for more points, and in particular it settles the triangular puzzle we set out to solve: if the corners of the triangle are at x, y and z, then the points inside the triangle are all those of the form $ax + by + cz$. where $a + b + c = 1$. This is the convex hull of the original set of points. By running through triplets of positive numbers a, b, c that add up to 1, a computer can quickly colour the convex hull green, producing the flat triangular panel we wanted.

In fact, exactly the same idea extends to larger collection of points too. In general, the convex hull of a set of points on the flat plane will be the shape formed by an elastic band, enclosing all the points as tightly as possible. (For points floating in three-dimensional space, we can think instead of an elastic bag hugging the points.)

Voronoi tilings and the mapping of disease

Besides being a mainstay of computer modelling, Delaunay triangulations have many applications beyond the computer sphere. Indeed, Delaunay originally invented the procedure to understand the intricate geometry of three-dimensional crystals. They also feature in mapmaking, where they manifest themselves as Voronoi tilings, named after Delaunay's teacher Georgy Voronoi. He first studied these configurations in 1908, but Voronoi tilings are now commonly used in a range of disciplines, from meteorology to the social sciences, enabling many seemingly difficult questions to be answered quickly.

They work in this fashion. Pick any arrangement of nodes on a map. There is a simple way in which this collection can then serve to divide the map into different regions, by adopting a rule that every location on the map is assigned to the node closest to it. The boundaries of the resulting Voronoi tiles – which can be of various sizes and shapes – occur at places that are *equal distances* between their two nearest nodes.

Where do triangles come in? It is not immediately obvious, but the geographer's Voronoi tiling is actually the same thing as a computer programmer's Delaunay triangulation, albeit somewhat disguised. More accurately, the two arrangements

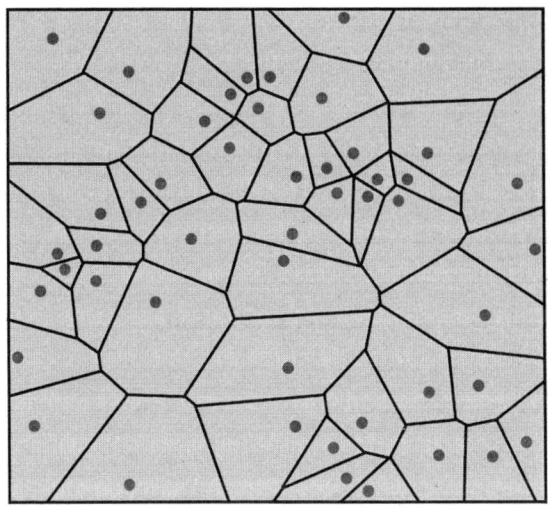

Any arrangement of nodes produces a corresponding Voronoi tiling.

are 'dual'. Starting with a set of nodes and the corresponding Voronoi tiling, if we then draw edges between those nodes whose Voronoi tiles touch, the result is a Delaunay triangulation.

As sometimes happens with useful and natural concepts, Voronoi tilings were being put to work long before Voronoi explicitly described them. A notable example occurred in the mid-19th century, at a time when many London households collected drinking water from public pumps. Needless to say, those charged with fetching the water usually chose the nearest pump, thereby dividing up the poorer parts of Victorian London into a Voronoi tiling. These observations took on an acute importance when, in 1854, the city experienced a severe outbreak of cholera, claiming over 600 lives. At the time, the bacterial basis of the disease was not understood, and the dominant theory was that it was spread by 'miasma', polluted air. When Dr John Snow plotted the cases of cholera on a map of London, he realized that the pattern corresponded closely to what was in effect one Voronoi tile, around the Broad Street water pump in London's Soho area. With this observation, he identified the infected pump as the source of the outbreak and convinced the authorities to shut it down. This work also marked the more general discovery that cholera is a water-borne disease.

In the years since John Snow's work, Voronoi tilings have been used to solve many other practical problems. Imagine a supermarket chain wishing to open a new out-of-town store. The company's planners may not want to be too near the competition or reduce the business at its existing stores, so – other factors being equal – might search for a location as far as possible from all other supermarkets. But how can they find this place? The first thing to do is to represent every existing supermarket by a node. The goal is to locate the *largest empty circle* that can be drawn on the map avoiding all the nodes. The centre of this circle would be the spot for the new shop.

Drawing the Voronoi tiling for this collection of nodes allows this position to be found easily. It is guaranteed to be at one of the vertices of the tiling, meaning a place where three (or more) tiles meet; more specifically, it will be at the vertex whose *nearest* Voronoi node is more distant than for any other vertex.

The limits of triangulation

The computer programmer's basic method of triangulation takes something for granted, namely that every two-dimensional surface has a triangulation. This seems to be a matter of common sense, and, comfortingly, it is at least true. Disconcertingly, however, the same assertion becomes false in higher dimensions.

Here, the equivalent of a triangulation is to break a shape into simplices. A simplex is a line segment (enclosed by two points), a triangle (formed from three points), a tetrahedron (four points) or suitable higher-dimensional equivalent. In three dimensions, the space will be divided up into tetrahedral shapes, and the equivalent of a Delaunay triangulation will then consider spheres instead of circles. To conform to the Delaunay principle in higher dimensions, the definition is that the sphere passing through the four corners of any tetrahedron should not contain any other nodes.

Higher-dimensional Delaunay triangulations have several applications in engineering: by replacing the original curved object with a triangulated one, many difficult problems become tractable, such as understanding the flow of heat through a material, or the tensile forces to which different parts of a material are subjected. But in 1982 the US mathematician Michael Freedman caused a shock, with the discovery of a four-dimensional shape which – remarkably – cannot be triangulated at all, due to its intrinsic knottedness.

MIRRORS AND MOLECULES

The (a)symmetries of
the universe

In our everyday lives, we are familiar with the experience of slipping up when giving or receiving directions – saying 'right' when we mean 'left', or vice versa. It is an embarrassing *faux pas* – after all, we learned our left from our right as children, didn't we? Yet there is still some mental processing involved in any particular situation. This need to pause for thought hints at the surprisingly rich questions about the nature of left and right: do they actually exist in any objective way, or are we simply referring to our environment as we individually perceive it? For two people facing each other, one person's left is the other's right, so what exactly is the difference between the two? Might there even be some universal notion of left and right, which could distinguish us from our reflection?

These may sound like idle speculations, but are anything but. They take us into a world of mirror universes and fundamental debates about symmetry. These have been major topics of scientific enquiry over the last hundred years, yielding some spectacular and unexpected discoveries. At the most profound level these are questions about the nature of the reality itself and whether it – and we – are subject to laws of geometrical symmetry.

Adventures in orienteering

In geometry, there is a fundamental distinction between orientable and unorientable spaces. An orientable space is somewhere where we can make a consistent choice of left and right, which remains valid across the whole space. This sounds straightforward, so the surprise is that *unorientable* spaces should exist. The most well-known example, enjoyed by several generations of mathematics students, is the Möbius strip: a simple rectangular strip of paper, twisted by 180° in the centre, with its ends then stuck together. To appreciate its strange unorientability we might imagine a two-dimensional person living in this surface – *in* it rather than *on* it, because to live on something implies a three-dimensionality that the Möbius strip does not possess. Suppose our flat friend decides to call one of his hands his 'right', and the other his 'left'. This seems unproblematic, and being dextrous he manages to bounce a ball with his right hand while waving to us with his left. However, the surprise comes when sets off on an expedition around the strip, bouncing and waving as he goes. By the time he comes home, his hands will have switched sides. He will be playing basketball with the left, and waving with the right. Just by going on a walk around the loop, the man has become his own mirror image – which is exactly the definition of an unorientable space.

Another unorientable space beloved of mathematicians is the so-called Klein bottle (see illustration in *The hole story*). Unlike the Möbius strip, but probably like our own universe, this shape has

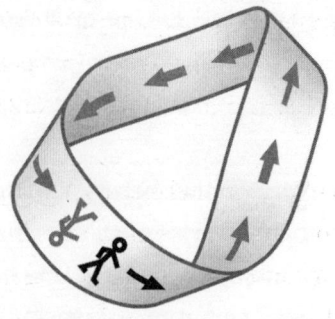

By walking around a Möbius strip, a 2-dimensional man will become his own mirror image.

no edge or boundary. One way to make one is to glue together two Möbius strips along their edges. The resulting space is internally coherent, but has the property that any version one tries to build will have to pass through itself somewhere.

The Möbius strip and Klein bottle are both two-dimensional spaces, meaning that each small region looks like a patch of flat plane. Their overall unorientability comes from the fact that the large-scale space is twisted around a hole in the third dimension (see *The hole story*). This hole would be invisible to an imagined inhabitant of the space, though, as we have seen, its existence might be deduced indirectly. But the presence of a hole is, on its own, no guarantee of unorientability: an inhabitant of a doughnut-shaped planet would find themselves in a perfectly orientable world.

For over a century, geometers have known of unorientable three-dimensional spaces too, and indeed of those in higher dimensions. This raises the largest question of all: what of our own universe? If it were not orientable, it would be possible, in principle, for a future astronaut to return from her expedition (perhaps having travelled around some sort of strange wormhole), apparently having developed *situs inversus*: the rare medical condition in which internal organs are reversed. Alternatively, if the four-dimensional space–time of our universe were unoriented, she might return to earth to find time running backwards.

It might be a relief, therefore, that the current consensus is that our universe is very likely to be orientable, as we shall see. That means we can at least make a coherent choice between left and right. But is it a free choice, or is there a correct answer? The real question here is, of course, not about language – we can easily imagine the meanings of 'left' and right' being transposed – but rather a conundrum in fundamental physics. We want to know whether one direction is preferred to the other by any physical processes. To put it another way, if the entire universe were to be replaced by its mirror image, could there be any way of telling?

It is worth catching up with our two-dimensional friend. Following his perambulations on the Möbius strip, let's imagine he's moved to a rather less exotic home: a single triangular room. If it is an asymmetrical triangle, he might be able to define 'left' as 'the direction in which the floor meets the wall at a sharper angle'. But if the triangle is symmetrical (isosceles), he cannot make this definition. In the asymmetrical case, if his triangular universe flipped while he was asleep, he would certainly notice when he woke up. In his isosceles universe he'd have a hard time spotting the difference.

The same applies in three-dimensions, which we can see if we incarcerate our friend as a prisoner in an empty cubic room, which has a yellow floor, one red wall, and the remaining walls and ceiling white (and indistinguishable). If the room were to flip during the night, our friend might not notice the difference on waking. This is because this room has reflectional symmetry. Flipping across a mirror through the centre of the red wall would leave the whole room looking the same. Flipping across a different mirror might move the red wall, in which case the prisoner might be momentarily disoriented on finding the red wall not where he expected. But even now, the reflection would be identical to a rotation of the room. So, after a moment or two to reorient himself, he would notice no difference in his environment from the day before.

It would be a different story, though, if two adjoining walls were painted red and green. Flipping across any mirror would mean that the green wall would be visibly on the 'wrong' side of the red one. Such a room has no reflectional symmetry, that is to say, it is impossible to reflect it while leaving it looking the same. Our friend would know something was up.

Therefore, to refine our central question, since both possibilities are geometrically possible, does our universe have reflectional symmetry or not? A way to tackle this is to move from the vastness of space down to the properties of molecules and individual atoms.

Chiral chemicals and particular poles

As with a room adorned with red and green walls, some chemicals come in both left-handed and right-handed varieties. The technical term for this is chirality, meaning 'handedness'. The simplest molecules are *achiral*, so have no handedness: to replace them with their mirror images would mean no change. A water molecule is one such, because a single oxygen atom sits between two hydrogen atoms, and the underlying symmetry means that, no matter how you reflect it, the result can then be rotated so as to be indistinguishable from its previous state.

Other molecules, though, are *chiral* and therefore without reflectional symmetry, and their mirror images are fundamentally different from the original. Louis Pasteur discovered the first one, tartaric acid, in 1848. Chemists call the original and mirror images the 'enantiomers' of the chemical. Many of the molecules of life, such as the amino acids that build up our proteins, come in two possible varieties, as does DNA itself (see *Untangling the stuff of life*) as well as the sugar molecules that our bodies use as fuel. In life on earth, the proteins from which we are constructed are largely left-handed. There is no reason, in principle, why elsewhere in the universe there may not be life-forms similar to those of earth but built from right-handed proteins instead; these aliens would be our mirror images on the molecular level.

Left- and right-handed versions of the same molecule can have different chemical properties. When these are not understood properly, the results can be serious. A notorious case in point was the drug Thalidomide, created by its German manufacturer during the late 1950s as an antidote for morning sickness in pregnant women. While its molecules' right-enantiomer did achieve this effect, it was realized too late that the left-enantiomer caused birth defects, and thousands of children were born with undeveloped limbs: what was not known, until tragically too late, was that one enantiomer can spontaneously convert into the other inside the human body.

Thalidomide demonstrated all too clearly the reality of chemical chirality. But chirality is also a *relative* phenomenon. The reason the two enantiomers react differently within the human body is that DNA is itself chiral. In an alien world, where life-forms have left-handed DNA rather than right, left-handed Thalidomide would be the dangerous form.

Using this terminology, then, is there any *absolute* sense in which the universe itself is chiral? For many years, physicists defined this question as the parity conservation problem, and most believed that its answer was 'no'. The prevailing thinking was that all the laws of physics would remain the same in a mirror-image universe. It was at the sub-atomic level – the minute world of protons, neutrons and electrons – where the discoveries were made that turned all this on its head.

Modern physicists believe that there are four fundamental forces of nature. Gravity keeps us on our planet and maintains its orbit around the sun, while electromagnetism keeps electrons orbiting the nuclei of atoms. The final two forces are only usually seen within the atomic nucleus itself. The so-called 'strong nuclear force' is what binds the nucleus together; the 'weak nuclear force' is what sometimes causes it to blow apart.

Of these four forces, the last has a unique property, as may be seen in the case of Cobalt-60, a radioactively unstable substance. The weak nuclear force is liable to cause a proton to spontaneously switch into a neutron within the nucleus, turning it into a nucleus of Nickel-60. During this process, an electron (known in this context as a beta particle) is emitted from the nucleus. The question, which Chien-Shiung Wu addressed in the United States in 1957, was in which *direction* this electron was fired. Her enquiries had been inspired by a theoretical paper by two Columbia University physicists, Tsung Dao Lee and Chen Ning Yang, who had noticed that, while the symmetry of gravity, electromagnetism and the strong nuclear force had all been well established, no-one had yet looked into the situation for the weak nuclear force. (They won the Nobel Prize for this in 1957.)

It was already known that the beta particle would emerge from one of the two poles – north or south – of the nuclei, meaning the two points that remain stationary as the nucleus spins. But which one? The poles may be distinguished by the direction of rotation: if observing the north pole from above, a nucleus seems to spin anticlockwise, while at the south pole, observed from below, it spins clockwise. In a mirror-image universe, the poles of the nucleus should be reversed.

Astonishingly, Wu's experiments showed categorically that electrons appear more often at the south pole than at the north, rather than equal numbers from each pole. This was a clear parity violation: in a mirror-image universe, therefore, more electrons would emerge from the north pole than the south – so there would be an observable difference. In the years since, further examples of parity violation have been found, all involving the weak nuclear force.

One consequence of Wu's remarkable discovery is that we can now be more certain that our universe represents an orientable space, one in which it is possible to make a *universally valid* choice of left and right. In beta decay, it seems likely that nature has indeed made such a choice. If our universe were unorientable, there would have to be another part of the universe where Cobalt-60 decays oppositely from the way it does on earth, with more electrons appearing at the north pole of a nucleus. Between these two regions, there would presumably have to be some sort of transition barrier, where the direction of beta decay switches.

While not beyond the bounds of possibility, the available evidence strongly suggests an orientable universe, supporting a universal law of beta decay.

Symmetry's last stand

The dramatic discovery of parity violation overturned millennia of intuition about the physical world. But all was not over for the advocates of symmetry. Some physicists fought back, and hypothesized that the combination of a left–right parity swap with a simultaneous reversal of all electric charges would amount to a true symmetry of the universe. However, this version – given the name CP-symmetry – fared no better than the original; in 1964, two American researchers, James Cronin and Val Fitch, found violations of that principle too, for which they later won a Nobel Prize.

Today there is one symmetry still standing: CPT-symmetry, standing for 'Charge, Parity and Time'. This holds that a simultaneous exchange of left and right with an inversion of all electric charge and the reversal of time should be a symmetry of the universe. So far, all known physical processes obey this law.

Symmetry's final fortress is not unchallenged, however. Plenty of experimentalists are lining up to test the strength of its walls, to the highest possible levels of precision. The good news for defenders of CPT-symmetry is that no breaches in its defences have yet been found, and there is some reason to hope that it may withstand all attacks. Whether or not it holds, we can at least have confidence that the geometry of symmetry will continue to enhance, and occasionally overturn, our understanding of the fundamental laws of our universe.

SYPHILIS AND CHRISTMAS LIGHTS

The mathematics of group-testing

In 1943, in the midst of the Second World War, officials from the US Army became concerned about the number of conscripted soldiers infected with syphilis. But how serious was the problem, and how could the affected men be winkled out? To answer this question would require investigations, but with blood tests expensive and time-consuming to carry out, the top brass wondered whether there might be some way to speed the process up. The question was picked up in an important paper by the statistician Robert Dorfman – and the highly practical mathematical subject of combinatorial group-testing was born.

Dorfman's new approach came from a single, simple idea. Instead of taking blood samples from each person and testing them individually, why not first mix the blood into groups? If the combined sample for 100 people tests negative, then all of them can be eliminated from the investigation, thus saving on 99 unnecessary individual tests. Of course, there is also a downside to this approach: if the combined sample were to come back positive, it might be that all 100 were carrying the disease – or just 1, or any number in between, and further tests would be needed.

But how many, and of what kind? Clearly, the desire would be to minimize time and effort to get to an accurate result.

The first factor to consider is the existing information about the prevalence of the disease. If we imagine there is just one infected person hiding in a total population of 128, testing everyone individually will certainly be a waste of resources. There is a much better way – dividing and conquering, whose earliest-known algorithm dates from around 200 BC. First, divide the group into two halves, say numbered 1–64 and 65–128, and test each group together. Then discard the negative half (say 65–128) and repeat the trick, now with numbers 1–32 and 33–64. This process of halving and discarding would eventually narrow the groups down to one single case, in a sequence of seven tests: 1–64, 33–64, 33–48, 33–40, 37–40, 37–38, 38.

The total pool of 128 is a convenient starting point, because being a power of 2 (in fact 2^7) it can repeatedly be divided in half and half again, exactly. Actually, the pattern can be seen more clearly when the numbers are written in the base-two number system (binary) rather than in our usual base-ten (decimal) system. Here numbers are expressed using only the symbols 0 and 1, known as 'bits' (or 'binary digits'; see *Avoiding bad language*). In the binary system, the number 38 is written as 0100110, and successive questions amount to a working through of the string of numbers, asking, in effect, 'What is the first bit?', 'What is the second bit?', etc.

This process of divide-and-conquer becomes more efficient the larger the pool. It requires only ten tests to identify a single infected individual from a pool of 1,024, and if that increases to 1,048,576 only twenty tests are needed. Of course, however, these examples rely on the fact that we had *a priori* knowledge that there was exactly *one* infected individual. When the exact quantity of infected cases is unknown – or more numerous – we may need to think differently. In fact, crude though it may appear to a mathematician seeking an elegant, swift solution, individual testing remains the best method when the rate of infection is known to be high. But if we have grounds to expect only a small infection rate of around 1 or 2 per cent, the approach of divide and conquer remains viable.

Combinatorial group-testing began in health assessment, but the same line of thought is equally valid across a range of applications, such as quality control in manufacturing processes. The only unalterable criterion is that groups rather than individuals must be testable. The classic example is a string of Christmas tree lights, where the whole string will work only if every individual bulb is functioning.

How many tests make it worthwhile?

The inevitable question at this stage, between our extremes of a single isolated infection and a mass infection, is: how do we decide what sort of testing methodology to follow? In an industrial setting, if we want to test 1,000 products, we might initially test them individually, or in groups of 2, 4, 8, 10, 50, 100, 250, 500, etc., but can we decide which is best?

Robert Dorfman's original analysis considered a two-phase process: first, divide the targeted population into groups; then, should a group test positive, test every individual within it. To decide on size of groups, the starting point for the analysis is a particular, critical number, which we might call p. This is the rate of infection or defectiveness, meaning that *the probability of a randomly selected item being defective* is p. (It may be estimated by doing some preliminary tests on randomly selected items.)

The rationale behind the whole process is to minimize the total number of tests. In a population of 1,000, if we initially test groups of size n, the number of groups will be:

$$\frac{1000}{n}$$

This, then, is the number of tests needed in the first phase of the process, so now the question is: how many of them are likely to test positive and therefore contain a defective item? We could call this likelihood q. This means the average number of positive groups will be q multiplied by the total number of groups:

$$q \times \frac{1000}{n}$$

With the first phase complete, the second is to test all the individuals in the positive groups. Since we can expect $q \times \frac{1000}{n}$ such groups, and each one contains exactly n items, multiplying these two numbers together tells us the number of individual tests needed during the second phase. This multiplication simplifies into the expression $1000q$. From this we can arrive at an expression for the total number of tests across the two phases:

$$\frac{1000}{n} + 1000q$$

Since our aim is to find the value of n that requires the smallest total number of tests, this means minimizing this expression. The mystery factor here is the number q, the odds that a single group tests positive. It must somehow be related to p (the chance of an individual testing positive) as well as n (the size of the group), but in what way?

By way of example, we might imagine that 1 per cent of the total population of items is defective, meaning $p = 0.01$. It follows that the probability of a single individual being healthy can be expressed as $1 - p = 0.99$. If we arbitrarily decide to test in groups of $n = 20$, probability theory now tells us how to calculate the chance that a group of 20 items tests negative. This simply requires each of the 20 items to be healthy, each of which has a probability of 0.99. So we need to multiply 20 lots of 0.99 together, making the chance that our group tests negative 0.99^{20}, which approximates to 0.818.

The other side of the coin is the probability that the group *does* contain at least one defective item, and so tests positive. This is exactly the number q we want. In this example therefore, q is approximately $1 - 0.818 = 0.182$. With a little algebra, we can generalize this to get a formula for q, in terms of p and n:

$$q = 1 - (1 - p)^n$$

All this preparatory work now answers the basic question as to how many tests will be needed. Using our existing values of $p = 0.01$ and $n = 20$, in our example the total number of tests would be approximately 232. Although this is undoubtedly a major saving on individual testing of 1,000 products, it is not the best possible result. Dorfman showed in his studies that when an infection rate is 1 per cent – that is, in our formula, $p = 0.01$ – the optimal group size is actually 11, which, when the formula is run, delivers a result of 196 tests. In fact, he presented some very revealing findings for different values of p:

Infection rate p	Optimal group size n	Total number of tests needed for a population of 1,000
0.1%	32	63
0.5%	15	139
1%	11	196
2%	8	274
5%	5	426
10%	4	594
20%	3	821
30%	3	911

One trend stands out. At an infection rate of 10 per cent, almost 600 tests are needed for the group of 1,000 people, and at a 30 per cent infection rate, requiring 911 tests, the exercise is scarcely worth the bother. The lesson is that, as the infection rate increases, the optimal group size shrinks, and the potential efficiency saving over individual testing diminishes. Once p exceeds 30 per cent, group-testing's usefulness ceases.

Improving the test

The story does not end with Dorfman; group-testing is an ongoing area of research. One obvious line of investigation is multi-phase testing. Instead of the two phases – groups followed by individuals – we might have a sequence of groups and subgroups of smaller sizes, until arriving at the level of individuals. In a sense, this returns us to the divide-and-conquer approach, but now the group sizes can be varied, as Dorfman did in the two-phase test.

There are subtler possibilities too. To take a very simplified example, if we have just three individuals to test (A, B and C), and we know that at most one of the three is infected, we can proceed as follows: test the pair A and B, and then the pair A and C. Together, these two tests will identify the infected individual, if there is one: if both tests are negative, all three are in the clear; if both tests are positive, we know A is infected; if one test comes back positive, we will be able to identify whether it is B or C.

In more complex cases this philosophy can bear rich fruit. What is more, this line of thinking forges a connection with the technological form of group-testing underlying the flow of digital information in the computer age: error-correcting codes (see *Avoiding bad language*).

Analysis along these lines can yield valuable information, including telling us the best we can possibly hope to expect to achieve in terms of minimizing the number of tests. Our divide-and-conquer algorithm needed just seven tests to identify a single infected individual from a population of $2^7 = 128$, and ten tests when the population was $2^{10} = 1,024$. The general rule here is that to identify a single case of infection (or defective item) from a population of n, the divide-and-conquer method needs at least $\log n$ tests, 'log' in this case meaning the logarithm calculated to base two – the number of twos need to be multiplied together. Which is just to say that $\log 128 = 7$ and $\log 1024 = 10$. This principle tells us that this is the best that any procedure can hope for. (For more on logarithms, see *The rise of* homo economicus.)

If there is more than one infected individual (or defective product), what happens to this best possible value? The answer is as follows: if there are d-many infections in a population of n, however you choose to proceed, the minimum number of tests needed will be:

$$d \times \log\frac{n}{d}$$

With a population of $n = 80$ containing $d = 5$ infections, the value of $\frac{n}{d} = 16$, and the absolute minimum number of tests needed will be $5 \times \log16 = 5 \times 4 = 20$.

From syphilis to DNA

As it happens, Dorfman's pioneering analysis in the context of potentially syphilitic servicemen was not put to use at the time. But since then his day has come, and his ideas have spread far and wide. Today, they are commonly applied across the health industry, being well suited to – for example – screening blood-donations for HIV and hepatitis, situations where the prevalence of infection is very low, making the efficiency savings considerable.

At the same time, group-testing is finding ever more applications in modern science. One example is the pharmaceutical industry, where the massive costs of developing new drugs and treatments are keen encouragement to finding new efficiencies in the testing process. When searching for, say, a new drug to kill a particular virus, typically the first phase is to try many thousands of candidate chemicals. This is well suited to group-testing, especially the early stages of investigations: the chemicals can be mixed in batches, and each mixture applied to a population of viruses. If a mixture has no effect, all the chemicals within it can be eliminated from future enquiries, while the components of an effective mixture can be tested further.

Group-testing is also at the forefront of the search to understand genetics and combat genetic factors in disease through DNA sequencing. In this field, researchers might be searching for tiny subsequences within a long strand of DNA offering millions of possibilities. By applying group-testing methodology and chopping the strand into shorter segments, grouping it and then applying the opposite subsequence to each group, researchers are able to bring the task into manageable proportions. If all the DNA groups are exposed to this dual subsequence, only those which contain the sought sequence will react. Those can then be tested again, segment by segment, while the rest are eliminated.

Blood screening, groundbreaking drugs and DNA sequencing (see *Untangling the stuff of life*) are all serious and worthy applications. On a lighter note, we can conclude with a famous puzzle, which illustrates the gains that group-testing can provide. Imagine we are shown ten large bags of coins. Nine of them are filled with genuine gold coins, weighing 10 grams each, but one contains only counterfeit coins, each weighing only 9 grams. How can we determine which bag contains the fakes, using only one measurement?

After a little head-scratching, if we've absorbed enough of the logic of group-testing, we will find an answer. First, number the bags 1 to 10. Then take one coin from bag 1, two from bag 2, three from bag 3, and so on. This gives a total of 55 coins, which we now weigh together – our one measurement – and this reading will tell us the answer. If bag 1 is guilty, the total weight will be 1 gram short, so read 549 grams. If it is bag 2, it will be 2 grams short, so weighing 548 grams; and so on. In general, if the total weight is $550 - n$ grams, then it is bag n that contains the fakes.

From the erudite to the entertaining, group-testing applications continue to expand from their Second World War beginnings.

THE CHAOS IN THE FISHPOND

The untidy growth of populations

In the late 18th century, the English economist Thomas Malthus issued what amounted to an apocalyptic warning. He wrote that 'population, when unchecked, increases in a geometrical ratio', as a consequence of which he could foresee a time when a multitudinous populace would so overwhelm available resources that humanity would be pushed back into a new dark age.

It is certainly the case that populations – whether of insects, plants or humans – are unlikely to remain constant over time, and for centuries before and after Malthus thinkers have explored the potential of mathematical methods for predicting and modelling population change. Today, ecology is one field where this thinking has had an important impact, and one doesn't have to look very far to realize why – from planning policy on how to manage the world's fish stocks to protecting the world's rarest species on land, precise measurements and reliable models are highly desirable goals.

The simplest approach – and therefore one of the most useful – involves equations to measure difference. The essential idea here is straightforward, but the consequences can be astonishingly complex, and even – in the full technical sense of the word – chaotic. (For more on chaos theory, see *One small step . . .*) Difference equations are found across the social and physical sciences, to model all manner of phenomena, with ecology one of their most prominent applications.

Fibonacci's rabbits and Tribonacci's threesomes

Before we try our hand at modelling natural phenomena, it is worth getting some familiarity with the mathematical underpinnings. Predicting growth has its mathematical roots as long ago as 200 BC in a celebrated sequence of numbers investigated by Indian scholars, who found it useful for describing the number of metres in lines of Sanskrit poetry. But the sequence takes its modern name from the Italian mathematician and traveller Leonardo Fibonacci, who studied it in depth, and introduced it to Europe in the early 13th century. His particular context was in estimating the growth of nature's notoriously fecund creatures – rabbits.

The Fibonacci sequence is an infinite list that runs 1, 1, 2, 3, 5, 8, 13, 21, etc. Its defining characteristic is that *each number is the sum of the previous two*. That rule can be expressed by the following equation:

$$x_{n+2} = x_n + x_{n+1}$$

where x_n stands for the number in the list. This law is an example of a difference equation, also known as a recurrence relation. Specifically, the Fibonacci rule is a 'second-order' difference equation, because each number is defined by the *two* that precede it. Naturally enough, a 'third-order' difference equation defines each number by the three preceding it. A related third-order equation is:

$$x_{n+3} = x_n + x_{n+1} + x_{n+2}$$

This law produces the sequence of so-called Tribonacci numbers, where each term is the total of the three before: 1, 1, 1, 3, 5, 9, 17, 31, etc.

First-order difference equations also exist, with each number determined solely by its immediate predecessor. For example, the equation $x_{n+1} = 2 \times x_n$ tells us that each number in the sequence is double the previous one. The obvious sequence obeying this rule is 1, 2, 4, 8, 16, 32, etc. This is the *geometric growth* of Malthus's nightmares: a population that grows by a fixed proportion year on year – in this case doubling. But here we reach a subtle point, because, on reflection, the sequence could equally well be 3, 6, 12, 24, 48, etc., or any other sequence of straightforward doubling.

The moral of this is that a difference equation *on its own* will not usually specify a unique sequence. That is true even for the Fibonacci equation itself; for example, the sequence 1, 2, 3, 5, 8, 13, 21, 34, etc. also satisfies the defining relationship that each number is the sum of the previous two. This particular sequence constitutes the Lucas numbers, named after the 19th-century mathematician François Lucas. As with the Fibonacci numbers, they have a remarkable tendency to appear in nature, for example as the number of petals on various plants.

In order to specify a unique sequence, additional information is needed beyond the equation, and this is typically supplied in the form of boundary conditions, which usually give the starting value. (This makes sense. In real life, you start with what you have now, before predicting how it might grow in future.) To fully encapsulate the Fibonacci sequence we actually need the first two terms (reflecting the fact that it is a second-order equation). So we kick off with the information that $x_1 = 1$ and $x_2 = 1$, and then by repeatedly applying the difference equation, we retrieve our sequence: 1, 1, 2, 3, 5, 8, 13, and so on.

In many situations, knowing the principle underlying the sequence is not going to deliver a quick answer to a question such as 'What is the hundredth number in the sequence?' (that is, 'What is the value of x_{100}?'). Is there a better way of doing it than starting with the value of x_1, and working out all 99 intermediate values? This is the central question in the theory of difference equations, and the answer is sometimes 'yes', and sometimes 'no'.

If we begin (again) with the boundary condition $x_1 = 1$ and adopt the difference equation $x_{n+1} = 3 \times x_n$, we would get the number sequence 1, 3, 9, 27, 81, etc. At each stage, the value is the result of multiplying together a suitable number of threes. This can be expressed algebraically as $x_n = 3^{n-1}$.

This formula presents an archetype for a *solution* to a difference equation. So now, if we want to know the hundredth number (x_{100}) in the sequence, we simply plug $n = 100$ into this expression, getting $x_{100} = 3^{99}$. (In fact, this has an enormous value of around 2×10^{47}, that is 2 followed by 47 zeroes, testament to Malthus's insights about geometric growth.)

Does the Fibonacci sequence also have a concise solution to arriving at any number in the sequence quickly? Pleasingly, it does. For those interested in the details, it is encapsulated in what is known as Binet's Formula, after the 19th-century French mathematician:

$$x_n = \frac{(1 + \sqrt{5})^n - (1 - \sqrt{5})^n}{2^n \times \sqrt{5}}$$

Plugging in a value of $n = 100$ tells us that the hundredth Fibonacci number is 354,224,848,179,261,915,075.

Other difference equations do not have such easily expressible solutions; chief among them are those which give rise to chaotic effects, as we shall see.

Biological realities and Beverton–Holt

Fibonacci deployed his eponymous sequence in his 1202 work *Liber abaci*, where he analysed the number of rabbits in a garden from month to month. He solved the problem on its own terms, but unfortunately the assumptions of his model were wildly unrealistic and of little predictive use. He assumed, for a start, that his rabbits were immortal, so that all new generations simply augmented the pre-existing ones.

To be of any real use to a self-respecting biologist or ecologist engaged in gauging populations, difference equations need to be rather more sophisticated. We might suppose that, resources permitting, the population of fish in a large lake will triple in size every year. On its own, this information suggests the difference equation:

$$x_{n+1} = 3 \times x_n$$

where x_n represents the number of fish in the nth year.

This is Malthusian geometric growth once again. But let's get real – after all, we are talking about fish in a finite expanse of water, and the trillions of fish predicted by this equation to appear by the 26th year are obviously never going to arrive. There will, inevitably, be a maximum number of fish that the lake's ecosystem can support, and as the lake starts to get full, fish reproduction is bound to slow down. For our purpose, we can take this limiting value to be 1 million.

Thankfully, in the 1950s Ray Beverton and Sidney Holt produced a model to bring some reality to this scenario. It began with the same equation as before, but multiplied by a new term at the end, thereby limiting the final result:

$$x_{n+1} = 3 \times x_n \times \left(\frac{1}{1 + 0.000002 \times x_n} \right)$$

It requires a little explication. With this model, if one year's value x_n of the population is a small number, then $0.000002 \times x_n$ – whose origins are explained below – will be tiny, and the part inside the brackets won't make much difference to the overall outcome. In other words, we're not very far off the original $x_{n+1} = 3 \times x_n$, with the population still almost tripling in size each year (or nearly so). If our lake has 1,000 fish one year, the following year the number will have risen to 2,994.

But the crafty nature of the Beverton–Holt model is that, once a population grows closer to the critical figure of 1 million, the story changes. Here, the value of $\left(\frac{1}{1 + 0.000002 \times x_n} \right)$ will approach one-third, which causes the rate of yearly growth to slow right down.

What happens, mathematically, when our lake holds 1 million fish? If we plug in a value of $x_n = 1,000,000$ the Beverton–Holt model says that the population in the next year will be:

$$x_{n+1} = 3 \times x_n \times \left(\frac{1}{3}\right)$$

The 3 and the $\frac{1}{3}$ neatly cancel out, giving us $x_{n+1} = x_n$. The population ceases to change at all at this point.

This fact allows mathematicians to dub 1 million a 'fixed point' of the system – and what is more it is an *attracting* fixed point, since, no matter what the starting population is (so long as it is not zero), the system will inevitably tend towards a final value of 1 million.

Pleasing though the Beverton–Holt equation might be, the curious reader might well wonder where the number 0.000002 was plucked from. The answer is that the model works with two pieces of data:

First, a number r that represents the unconstrained rate of growth of the population: in the fish example, $r = 3$.

Second, the so-called carrying capacity K of the habitat, which was 1 million for the lake.

From these two pieces of data, a third number may be defined:

$$s = \frac{r-1}{K}$$

and that gives $s = 0.000002$ for our fish.

The general formula for the Beverton–Holt model is:

$$x_{n+1} + rx_n \left(\frac{1}{1 + sx_n}\right)$$

When broken down, this is a comparatively simple formula, and it has been successfully deployed in multiple real biological and ecological scenarios. What is more, as with the Fibonacci sequence, it has the major mathematical advantage of admitting an exact solution: a single formula that can give us the population of the lake (x_n) in any year.

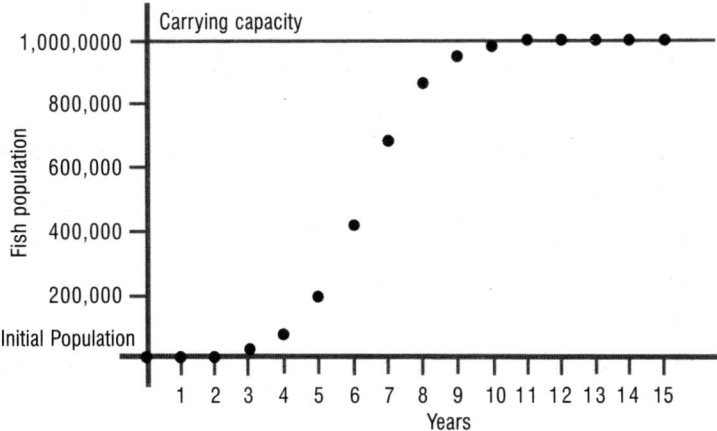

The growth of a population of fish, as predicted by the Beverton-Holt equation.

From fixed points to cycles and beyond

The 1 million fish in our example provided a *single* attracting fixed point, which is the simplest long-term outcome for a difference equation. Other possibilities are cycles of various length. For example, if one were to decide that the appropriate equation in some scenario was $x_{n+1} = 100 - x_n$, and we began with a starting value of $x_1 = 30$, then the result would be an interminable alternating sequence running 30, 70, 30, 70, 30, 70, and so on. This is known as a '2-cycle'. Other systems can produce cycles of greater length.

As with the fixed point in the Beverton–Holt model, these cycles will often be attractors, meaning that whatever the starting value of the sequence, it will eventually home in on the repeating cycle, and there is a comforting feeling of regularity. But not all difference equations are so predictable.

The Beverton–Holt model describes an isolated species. It does not take into account dynamics such as those between predator and prey. This can be a subtle business: if there are few predators, then the prey population will rise, which will cause the number of predators to grow next year. But with a lot of predators around, the number of prey will be reduced, in turn having a negative impact on the predator population. These ideas, which are so fundamental to the workings of the natural world, can be captured mathematically by a *system* of difference equations.

One of the more straightforward approaches is the Neubert–Kot model of 1992. Here we have two quantities that vary: x_n describes the density of the prey, and y_n that of predators. Two other fixed numbers are involved, r and s. The relationships between them are articulated by a somewhat complex-looking duo of equations:

$$x_{n+1} = (r + 1)x_n - rx_n^2 - sx_ny_n$$
$$y_{n+1} = sx_ny_n$$

This model allows a detailed analysis of when the predator–prey situation is stable, and when the two species drive one another towards extinction. What is more, as with many difference equations this model can be 'chaotic' (depending on the precise values of r and s), in the sense that, however you long you leave it, the populations will continually jump up and down, never settling into a stationary level (fixed point), or even a predictable, repetitive cycle. The differences over time may appear to be just random fluctuations. Biologically, this is a hugely important insight, as it contradicts the naive idea of an essentially static ecosystem.

From fishponds to fractals

We have looked at some of the more straightforward examples of difference equations, but even seemingly simple scenarios can give rise to chaotic effects. Magnificent images of fractals – those weirdly self-replicating but irregular-seeming patterns – often accompany this phenomenon.

One example can be generated by the ostensibly simple rule 'multiply the number by itself, and then subtract 1'. As a difference equation, this can be written as

$$x_{n+1} = x_n^2 - 1$$

Putting in the boundary condition $x_1 = 1$ gives rise to the sequence: 1, 0, −1, 0, −1, 0, −1, etc., producing a cycle that flickers endlessly back and forth between 0 and −1. But begin it with a with a boundary condition of $x_1 = 2$ and the equation produces a very different sequence: 2, 3, 8, 63, 3,968, 15,745,023, etc., growing without limit.

A tantalizing question, motivated by our concerns with population growth is: which starting values of x_1 produce a sequence that is bounded as opposed to one that outgrows any limits we might try to impose? The collection of all starting values for which the sequence remains bounded is called the equation's 'Julia set', in honour of one of the pioneers of fractal geometry, the mathematician Gaston Julia (1893–1978), who did much of his best work while lying in a hospital bed recovering from wounds in the First World War. When plotted on a graph, and allowing what mathematicians call 'complex numbers' as starting points, Julia sets produce some of the most astonishingly intricate images in the whole of mathematics.

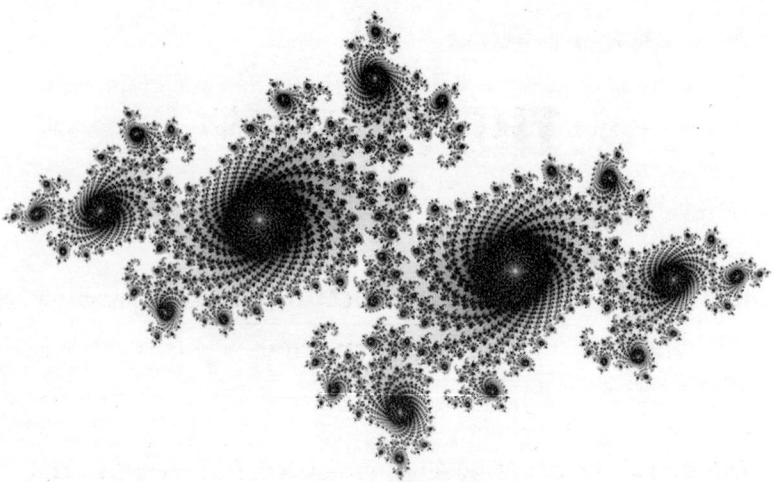

Julia sets exhibit characteristically inticate fractal patterns.

In recent years, Julia sets have been adopted by visual artists seeking mathematical inspiration. These now ubiquitous fractal images are perhaps the most obvious impact equations on the human world. But their importance runs much deeper. By providing mathematical models for articulating change over time, difference equations have given ecologists – and scientists from chemistry and physics to computer science – some serious tools of their trades.

THE RISE OF
HOMO ECONOMICUS

The mathematical basis of
decision theory

Our lives are beset with all-kinds of decision, from the trivial and mundane to the life-changing and even – for those with power and influence – world-changing. Easy decisions involve rejecting something disadvantageous in all respects for something roundly and palpably beneficial. But most decisions are not so straightforward, and most choices present both potential benefits and costs that must be weighed up and considered. Can mathematics in some way help make a correct decision, or at least provide a basis for comparing possible decisions?

Perhaps surprisingly, the answer is 'yes'. Indeed, a whole subject has developed to answer this question: decision theory, in which the most important notion here is the *expectation* associated with a choice. The basic idea is not too complex, yet decision theory can sometimes throw up some very strange and counterintuitive results. Many of the examples in this field revolve around the choices made in hypothetical games, since decision theory is closely related to another mathematical genre, that of game theory.

Horizons of expectation

It makes sense to begin with the assumption that we wish to make choices that are, in some rigorous sense, the very best possible: we are 'rational agents' rather than masochists or impulsive flakes. We could consider an example in which we are offered a choice to play one of two games. In the first, we toss a coin and if it lands head-up, we win £2, but if it's tails we win nothing. In the second, we roll a six-sided die, and if we roll a 6 we win £12 but otherwise we win nothing. Which to choose? The standard approach to questions like this is to compute the *expectation* of each game.

With the coin, there are two possible outcomes, each with its own reward and its own probability: heads has a probability of $\frac{1}{2}$ and a reward of £2, while tails also has a probability of $\frac{1}{2}$ but a reward of £0. The game's so-called expected reward comes from multiplying each reward by its probability, and adding up the results:

$$\left(\tfrac{1}{2} \times £2\right) + \left(\tfrac{1}{2} \times £0\right) = £1$$

An expected reward is not *literally* winnable – for the rules of the game do not allow winnings of £1. But it reflects the average winnings per toss, over a typical run of wins and losses.

Meanwhile the dice game also has two possible outcomes: a six, with probability of $\frac{1}{6}$ and reward of £12, and any other of the five numbers with the probability of $\frac{5}{6}$ and no reward. So, for this game the expected reward may be computed as:

$$\left(\tfrac{1}{6} \times £12\right) + \left(\tfrac{5}{6} \times £0\right) = £2.$$

Which game is therefore the most profitable? A rational player, almost by definition, will always opt for the larger expectation. But this is not how humans approach such decisions. People tend to be more risk averse, and of course are entitled to prefer a reasonable shot at a modest reward to a punt on hitting the jackpot. Nevertheless, there is a scenario where the rule holds good: if you had to pick one game and play it many times, you would undoubtedly be better off with the dice game in the long run.

The better pay-off for the dice game derives from what mathematicians call the law of large numbers, a central principle in probability, decision theory and economics. Essentially, the law states that in a game repeated often enough (say *n* times), the average reward – more specifically, the 'mean', i.e. total winnings divided by the number of games played – will come ever closer to the game's expected reward. (For more on the mean and other definitions of 'average', see *The law of averages*.)

In terms of algebra, if X_1 is the first game, X_2 is the second, and so on, then as the number of games played (n) increases, the closer the average gets to the expected value of a single game $E(X)$, signified in this expression by an arrow:

$$\frac{X_1 + X_2 + \cdots + X_n}{n} \rightarrow E(X)$$

Seasoned gamblers will know this. For example, when playing roulette, the expected number of red and black numbers is equal (so long as the wheel is fair). So the law of large numbers guarantees that over a long enough timescale, the proportion of red and black spins will be the same. But this emphatically does *not* translate into short-term predictions along the lines of 'we've just had four black spins, so red must now be due'. Over-optimistic punters are prey to this misconception, hence its colloquial name: the gambler's fallacy. (For more on expected values, see *Tulip bubbles and hedge funds*.)

Paradoxical St Petersburg

An understanding of the laws of expectation can therefore aid the dedicated gambler; more importantly it is a prized tool among statisticians and economists. Nevertheless, it can throw out very strange results. A notorious example is the St Petersburg Paradox, dreamt up by Nicolas and Daniel Bernoulli in the 18th century, which also relates to around the tossing of coins. In this game, the coin is tossed repeatedly until tails appears. What is more, our winnings increase, the longer the run of heads. So we win £1 if tails appears on the first toss, £2 if it comes on the second, £4 on the third, and so on. The reward doubling each time. There is a catch, however: you have to pay each time you play. The question is: what is a fair price for entry to the game?

The standard approach would be to calculate the game's expected reward. If that exceeds the entry price, then rationally you should agree to play, and if not, then not. This sounds reasonable, until we actually try to perform the calculation, when something very strange happens.

Each possible outcome has its own probability and reward: the probability that the game will end after one throw is $\frac{1}{2}$, and this case, we would win £1. There is a probability of $\frac{1}{4}$ that it will end after two throws, netting us £2. The chance that it will end after three throws, winning us £4, is $\frac{1}{8}$, and so on. Putting all these together, the expectation seems to come out as

$$\left(\tfrac{1}{2} \times £1\right) + \left(\tfrac{1}{4} \times £2\right) + \left(\tfrac{1}{8} \times £4\right) + \left(\tfrac{1}{16} \times £8\right) + \cdots$$

That is to say £0.5 + £0.5 + £0.5 + £0.5 + ... In other words, the 'expected' reward for this game is actually infinite!

This seems like a paradox. How can a single game be infinitely profitable? Yet there is a sense in which it is correct. Again, it comes from repeated playing. Surprisingly, if you play the game often enough, then *any* entry price will indeed eventually represent good value. The catch is the number of rounds of the game needed. If the cost is £10 per game, you are likely to have to play over a million times before coming out in profit. If the cost is £100 per game, the lifetime of the universe is unlikely to be long enough.

Monty Hall and paradoxical envelopes

Perhaps the most famous example of decision theory in action is the so-called Monty Hall Problem, named after the presenter of the US TV show *Let's Make a Deal*. Here, we are faced with three doors, labelled A, B and C, behind each of which is a prize. One of the prizes is an all-expenses-paid holiday to a destination of your choice; the other two are free copies of a book about the applications of mathematics. Needless to say, any sensible player will be hoping for the holiday, but with no information to go on, we must just pick at random – say Door A.

Now, before the host unveils our prize, he gives us an opportunity to change our minds. He doesn't tell us what is behind Door A, and neither will he reveal the holiday. But with a flourish he reveals that behind Door C is a copy of *Chaotic Fishponds and Mirror Universes*. He then offers us a choice: to stick with our original Door A, or swap to Door B.

Almost everyone's immediate reaction would be that it can

make no difference. After all it is surely now a straight 50–50 choice ... or is it? At this stage, sheer inertia means that most people do not swap, since they do not see it as increasing their vacation chances in any way.

The shock is, however, that inertia would be wrong. The *rational* move is to swap. After all, by not swapping we can only win if we picked correctly to start with, and the probability of that happening was $\frac{1}{3}$. It is *more likely* – with a $\frac{2}{3}$ probability – that our original choice was wrong. Given that likelihood, together with the new information that Door C conceals a book, the odds on Door B concealing the holiday have risen. Now, by swapping, we take advantage of that information and increase our chances of winning from $\frac{1}{3}$ to $\frac{2}{3}$. Put another way, if a player's original choice was a book (and the odds that it were are $\frac{2}{3}$), then by changing doors they will be guaranteed to win.

Many people have been sceptical – incredulous even – about this solution to the Monty Hall Problem but the theory is easily testable, with two people and three playing cards to represent the prizes. Indeed, this experiment has been run many times at schools around the world, and the results are conclusive: swapping is the more advantageous move.

While the Monty Hall Problem is baffling to many, the so-called Two Envelope Paradox, proposed by J.E. Littlewood and Maurice Kraitchik in 1953, remains mysterious even to experts. It involves two people – say Alice and Bob. Alice offers Bob a choice of two sealed envelopes, each containing a sum of money, with one containing exactly double the amount of the other. With no information to help him, Bob chooses at random. But before he opens the chosen envelope, Alice gives him the opportunity to change his mind. Should he swap envelopes?

The obvious – and surely correct – answer is that he has exactly a 50–50 chance of picking the larger sum, and swapping will make no difference. Instead though, Bob reasons this way: suppose the amount of money he has in his current envelope is A. Then the other contains either half or double it – $0.5A$ or $2A$ – with an equal chance of each outcome. So, he calculates the expected value: $(0.5A + 2A) \div 2 = 1.25A$. This analysis suggests that the rational thing for Bob to do is swap.

From a perspective of common sense, this is surely silly. But it gets worse: once Bob has swapped, he can reason in the same way again, which will lead him to swap back. The theory here seems to suggest that Bob's rational course of action is to keep swapping the envelopes back and forth indefinitely!

This problem continues to be hotly debated by today's mathematicians and philosophers, and poses genuinely difficult questions about the foundations of economics and decision theory. Several proposed resolutions have been suggested, which revolve on understanding the probabilities behind the sums of money in the envelopes. But the problem is considered by many people still to be open, and has spawned a large number of descendants – for example, we can amend the rules to allow Bob to open the envelope before deciding whether to swap. Does this make any difference? Meanwhile the issues it throws up are of ever broader relevance, even as far afield as quantum physics, where our efforts to grapple with the randomness at the heart of nature have given a new importance to understanding the limitations of standard tools for addressing expectation.

Utility and diminishing returns

Underlying decision theory and economics is the assumption that a rational agent wishes to increase a particular quantity, known as utility, which is some form of numerical measure of the benefit or disadvantage gained from a particular event or transaction: money is the obvious example. But it is a fact of life that the perceived value of any amount of money is unlikely to be inherent or fixed – to some extent it is in the eye of the beholder. Who could doubt that a windfall of £1,000 would be worth more to a poor man than to a millionaire? This phenomenon is known as 'diminishing returns', and one way mathematicians grapple with it is to assign utility to a sum of money not according to its face-value but its logarithm.

Logarithms (or 'logs') – invented by John Napier (1550–1617) – came about as a means of simplifying calculations involving multiplication/division of large numbers by converting them into problems involving only addition and subtraction. The logarithm of any number is that number expressed as a power of the particular base number that we are working with. For the decimal system (with a base-ten), the number $100 = 10^2$, and so we say that the log of 100 is 2. With a different fixed base, say base-two (binary system), we could say that the log of number 8 is 3, because 8 is 2^3.

With a base-two system, to find the logarithm of 128 we are effectively asking: 'how many 2s are multiplied together to produce 128?', and the answer is 7. If we double 128 and find the log of 256, the answer would then be 8. These examples illustrate an important fact about logarithmic growth: it is subject to the aforementioned diminishing returns. The higher the value of x

rises, the slower log x grows. If we start with a value of $x = 2$ then $\log x = 1$. Increasing x by just 2 to 4 raises the value of $\log x$ to 2. However, we have just seen that to lift the value of $\log x$ from 7 to 8 requires an increase in x of over 100.

Daniel Bernoulli returned to the baffling St Petersburg Paradox, taking the utility of the pay-off to be the logarithm of the prize-money. In doing so, the 'paradox' disappeared. Here, the expected utility from playing the game becomes

$$\left(\tfrac{1}{2} \times \log 1\right) + \left(\tfrac{1}{4} \times \log 2\right) + \left(\tfrac{1}{8} \times \log 4\right) + \left(\tfrac{1}{16} \times \log 8\right) + \cdots$$

Unlike the original calculation, this does not grow without limit. As it turns out, its value is log2. So any entrance price that is less than this represents good value to someone valuing things on a logarithmic scale. (Bernoulli also adapted this calculation to take into account the players' previous wealth.)

Homo sapiens versus *homo economicus*

The branch of decision theory surveyed so far considers how perfectly rational agents *should* behave. As we have seen, at the heart of this subject is the notion of expectation, and mathematicians consider the various ways in which a rational agent might consistently assign a utility function to a range of outcomes. The face-value and logarithmic utility functions are just two possibilities. Certainly, this work can tell us a great deal about how to make good decisions on a rational basis. But it cannot be the whole story.

The second branch of decision theory investigates how human beings actually *do* decide such matters. Over the last few decades, psychologists have carried out numerous experiments in this area, and the results testify to the human tendency often *not* to choose the 'rational' solution. For example, across many aspects of life most of us are very loss averse: we dislike losses far more than we enjoy gains.

We can return to tossing coins to see this natural caution at work. Suppose we are offered a game whereby heads loses us £100 while tails wins a little more, £110. Computation of the expected reward would indicate an overall profit:

$$\left(\tfrac{1}{2} \times -£100\right) + \left(\tfrac{1}{2} \times £110\right) = £5$$

So a rational agent would opt to play. Yet studies have shown that the overwhelming majority of people will turn down this game, no matter how rich they are: the thought of losing £100 is too painful to make the potential prize of £110 worth it. In 1997, Matthew Rabin showed that if a rational agent's utility function satisfies some very mild mathematical assumptions, and if that agent also turns down this game no matter what its current wealth is (as most humans will), then it is logically compelled also to turn down other games that most of us would jump at, so long as we are wealthy enough to afford it, such as heads loses £1,000 while tails wins £1 million. In fact, following the rational-agent logic, a game with a £1,000 loss must be rejected for *any* winning sum!

Psychologically, real people tend to fixate on *changes* in their wealth: the pleasure of profit and (especially) the pain of loss, while the rational agents make judgements about their *total* net wealth after winning or losing. Those decisions where a rational agent – sometimes dubbed *homo economicus* – diverges from human psychology can sometimes draw attention to poor decision making. In the above example, so long as you are rich enough that the loss of £100 will not cause you real difficulties, you should probably play the £100/£110 game.

Here is another example. We are offered two bets, each based on the roll of two dice.

In bet A, there is a chance of $\frac{11}{36}$ of winning £1,600, and $\frac{25}{36}$ of losing £150.

In bet B, the odds are $\frac{35}{36}$ of winning £400, and $\frac{1}{36}$ of losing £100.

Which to play? Experiments by the psychologist Sarah Lichtenstein tell us that most people prefer the safety of bet B to the chance of riches in bet A. But now here is another question: imagine that we have secured the rights to play one of the bets, but someone – call him Claude – comes along and asks to buy these rights from us. At what price do we sell?

A common reaction, influenced by the large difference in the winning sums on offer, is to value bet A more highly than B. Suppose we assign a value of £500 to A and £350 to B. This would be a fairly typical response, but it is also a flagrant contradiction of the previous preference for playing B over A. This phenomenon is what is known as 'preference reversal', and it is something that human beings are subject to, while *homo economicus*, of course, is not. To see the problem, imagine that Claude offers us the opportunity to buy the right to bet A for £400. Since we've just valued it at £500, we might reasonably think that this represents good value. Next, Claude offers us the chance to swap to bet B. Since we have already decided that we really prefer B anyway, we should certainly be happy swapping. Our downfall is nearly complete, because Claude's final move is to buy bet B off us at our agreed price of £350, leaving us £50 down with nothing to show for it.

If a picture is emerging of human beings as falling some way short of the rational model, that is, in a sense, true. But of course it is not the whole story. If *homo sapiens* has an instinct for caution, even one that sometimes misjudges the balance of risk, there may be good evolutionary reasons for this. On the African savannah, it is quite plausible that our forebears found it more important to guard against loss than to spot opportunities for profit, and this default tendency has been passed down to us. What we have that they did not are the mathematical tools to understand and analyse the underlying issues.

And as we have seen, the techniques of decision theory can illuminate human psychology, enhance economic understanding and help people towards making better-informed choices.

THE HOLE STORY

The shapes of the universe

Voids, chasms, and abysses – such phenomena have long exerted a pull on the human imagination. Holes seem to resonate with us, from the dizzying sense of standing on the edge of the world that one gets at the Grand Canyon, to Alice's surreal experiences down the rabbit hole. Metaphorically, we speak of emotional holes being left in our lives by bereavement. But this is not only a human idiosyncrasy, perhaps an evolutionary hangover from our days as cave-dwellers. The truth is that holes are profound and important objects, with the most famous examples being the black hole, sucking into oblivion all that strays into its path.

This fascination has also found an outlet mathematically, where curiosity about shapes and the possible holes they may contain has given rise to an entire subject, topology, emerging from radical new approaches to geometry in the late 19th century. In the 20th and 21st centuries, topology has provided tools for physicists to tackle questions of fundamental importance about our own world and beyond, from analysing the structure of ice to theorizing about the beginning and subsequent expansion of the universe.

But we should begin at the beginning. How many holes does a shape contain? It is a less obvious question than it sounds.

One piece or two, one cut or two?

Perhaps the simplest observation anyone can make about a shape is whether it comes in one piece or more. In technical terms, a shape comprising just a single piece is 'connected', a description that applies to triangles, spheres, straight-line segments, pyramids, the bagel-like torus and many other more exotic shapes. But if we were to take two triangles with a gap between them and treat the pair as a single object, the result would be a disconnected shape.

So far, so simple, but it is not always a completely obvious distinction. For instance, two interlocking metal hoops are *disconnected* from a mathematical viewpoint, since they form two separate pieces with clear space between them. The same is true for the classic Borromean rings, where none of their constituent three hoops passes through any other, but nevertheless the three cannot be separated. Yet if one ring is removed, the remaining two will automatically fall apart.

The story deepens when one introduces the subtly different concept of a shape being 'path-connected', which means that

The Borromean rings: removing any one of the three causes the remaining two to separate.

any two points on the shape can always be joined with a path of finite length, whether straight or curved. It seems self-evident that any path-connected space must also then be, in mathematical terms, connected, for a path cannot cross a gap. The surprise is that the opposite is not true. There are some mathematical spaces that are connected, but *not* path-connected.

A famous example is the topologist's sine wave. This unusual object comes in two parts: a vertical line ('the wall'), and a curved part ('the curve') which becomes ever more wiggly as it bunches up to the wall. (Technically, the curve is given by the equation $y = \sin\frac{1}{x}$ while the wall is the straight line $x = 0$.) Crucially, there is no way to divide the wall from the curve; there is not even a sliver of empty space between the two, so together the two form a single connected entity, but not one that is path-connected. If we attempt to trace out a path from a point on the curve, it will never reach the wall. The wiggles of the curve conceal its infinite length, which no finite path can ever traverse.

Disconnected spaces reveal very clear 'holes' in the sense of gaps between their constituent parts. By contrast, a wedding ring – certainly a connected shape – exhibits a hole of a different type, and of a kind harder to describe precisely. One approach to

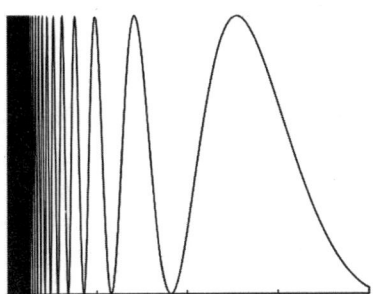

The topologist's sine wave falls between two definitions of connectedness.

capturing it mathematically is to ask how many times the shape can be cut before it becomes disconnected. Cutting across a ring once still leaves a single piece of metal, which might be pulled into a straight rod, and which remains connected; but a second cut will divide this rod into two. So the 'cutting number' associated with a wedding ring is 1.

A hollow torus, such as an inflatable rubber ring, tells a different story. One cut across it will turn it into a tube, open at each end. But this time a second cut need not divide the shape into two pieces, for one could slit along the *length* of the tube on one side, after which it would be possible to unroll the shape into a single flat sheet, which remains connected. But any third cut will disconnect it. So the torus's cutting number is 2.

A shape's cutting number is also known as its first Betti number (or B_1), named after Enrico Betti who tackled the subject in the early 1870s. There is also a more primitive zeroth Betti number (B_0), which simply counts the number of connected parts to the shape. For the Borromean rings $B_0 = 3$ but for any connected shape, by definition $B_0 = 1$.

Betti numbers of higher order $(B_2, B_3, B_4,$ etc.) count the number of holes of different kinds. For instance, an inflatable beachball comprises a single piece, and has no wedding-ring type holes either. Yet it does contain a hole of another type: the three-dimensional cavity that is filled with air as the ball is inflated. A marble placed within this cavity will roll around inside it but not be able to escape without puncturing the surface. As it has only one such void, the second Betti number (B_2) of this shape is 1. The same is true for the inflatable ring, and for a double-torus, which resembles an inflatable figure-of-eight.

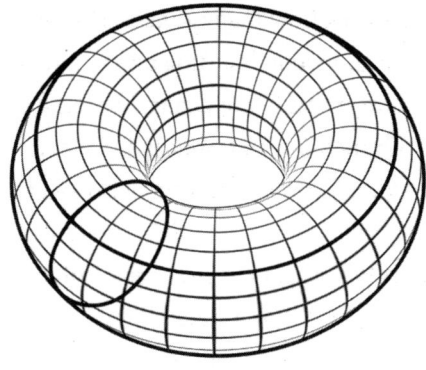

A torus. Cutting the torus twice causes it to be unrolled into a square.

Laying out our Betti numbers for connected parts, cutting number and cavities, we can begin to compare some properties of shapes:

	B_0	B_1	B_2
Straight rod	1	0	0
Solid ring	1	1	0
Hollow sphere	1	1	1
Hollow torus	1	2	1
Hollow double-torus	1	3	1

As we move beyond our usual realm into higher-dimensional spaces, Betti numbers of still larger degree are needed to describe the higher-dimensional voids we may find. A hypersphere, for instance, is a shape whose three-dimensional structure curves around to meet itself, just as happens for the two-dimensional surface of an ordinary sphere. Its Betti numbers are $B_0 = 1$, $B_1 = 0$, $B_2 = 0$, $B_3 = 0$ and $B_4 = 1$.

The Betti numbers have been popular, concise and efficient ways to count a shape's holes since Henri Poincaré proved in 1895 that they are topological invariants. This means that a shape's Betti numbers are unaltered by any amount of pulling, stretching or twisting – so long as the shape is not broken – making them highly robust and useful pieces of data. Such definitions go a long way to helping topologists categorize shapes together that, to the casual observer, might seem utterly different, while drawing formal distinctions between shapes that may superficially look similar.

The mysterious case of the disappearing hole

Their usefulness notwithstanding, Betti numbers do not account for some profoundly unexpected properties of holes, which may be revealed using the idea of a contracting loop. We can appreciate this by imagining a loop drawn on the surface of a sphere, such as our beachball: if we imagine that, once drawn, the loop can slide around the surface then we would be able to retract the loop gradually to a single point. The same trick might not, though, work on our inflatable rubber ring. It may be that the loop gets stuck around the circular hole in the centre. In fact, in these terms it may be said that the torus has *two* such circular holes, meaning two fundamentally different ways of drawing an uncontractable loop: one around the entire ring, and one around the cylindrical tube. This matches up nicely with the fact that $B_1 = 2$.

The surface of one of topology's favourite shapes, the Klein bottle described by Felix Klein in 1882, has a hole of a more unusual kind. If we draw a loop around it, from the top of the opening, around the outside, and then back up through the funnel, the result cannot be contracted to a point. This is proof positive that there really is a hole there. However, the strangeness

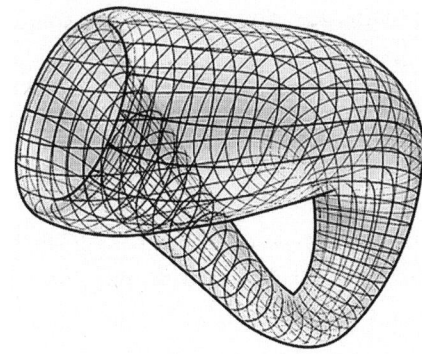

A Klein bottle exhibits two kinds of topological holes, one across the bottle's waist and one around its length.

of this hole means that it is not picked up by the shape's Betti numbers: as expected, $B_0 = 1$, reflecting the fact that it comes in one piece, and $B_2 = 0$, which tells us that the shape does not contain a three-dimensional cavity (if you play around with a Klein bottle, you will quickly realize that any marble placed inside will be able to roll out again). The shock comes at the first Betti number: $B_1 = 1$, rather than 2, as might be expected. This successfully identifies one ordinary circular hole: a loop around the waist of the bottle cannot ever contract but can only slide back and forth between the body of the bottle and the 'handle'. But this number misses out the loop that runs around the bottle, along the body and back along the funnel. The reason is that if we *double* that loop, by running around it twice instead of once, the resulting longer loop simply vanishes. That is to say, the doubled loop can now shrink away to a point.

It seems to defy common sense that a loop should be uncontractable while its double *does* contract. The reason is related to a famous fact about a Möbius strip: if you draw a line along its centre, it will only meet up with itself after traversing the loop twice. A Klein bottle is built from two such strips glued together.

This discussion all goes to show the extent to which holes have hidden depths. Indeed, they represent some of the most complex topics in modern mathematics, yielding profound applications to physics. Not the least of these are what physicists call 'phase transitions' between states of matter.

The matter of matter

In ordinary circumstances matter exists in three basic states – solid, liquid and gas. At ultra-high temperatures though, a fourth appears: ionized plasma. Solids and liquids are known as condensed matter, and here finer gradations are possible; For example, the liquid crystals used in digital displays exhibit properties of both liquids and crystalline solids. The positions of the molecules reflect the overall symmetry and order we would expect from a crystal including their neat arrangement in layers, yet these layers can slide across one another, causing the matter to flow as a liquid.

Another example is ice. We usually think of ice as a single phenomenon, but water may adopt no less than 15 different solid crystals (known so far), which vary according to how exactly the atoms are arranged. On earth, the commonest by far is referred to as I_h, where the molecules are arranged in layers of hexagonal rings. Yet in the earth's upper atmosphere another arrangement is occasionally found, I_c, where these layers are pulled together into a cubic arrangement, similar to a diamond. To further complicate the picture, most of the water in the universe is in none of these 15 crystalline forms, but is rather *amorphous*, in that the molecules are arranged haphazardly as happens in glass, rather than according to any crystal structure.

Under certain conditions, condensed matter spontaneously converts from one form to another. For instance, in 2009, 'Ice XV' was discovered, in which molecules are arranged into cells resembling slightly squashed, tilted cubes. This was achieved by cooling water to $-143°$ Celsius under huge pressure (nearly 10,000 atmospheres). These phase transitions are complex processes and pose real challenges to our understanding of condensed matter. The same ideas that apply to ice and liquid crystals are also relevant to questions of fundamental physics, including investigation into black holes, neutron stars, as well as the early universe, where the huge temperatures and pressures packed matter together in ways that rapidly changed as the universe expanded. And here topology can help.

Topological defects and cosmic textures

A highly fruitful avenue of research into phase transitions has involved the study of 'topological defects' that may emerge in the matter investigated. They resemble holes of different types, which can be analysed mathematically; but they are not *literally* holes, since typically they involve changes in structure rather than empty spaces. Yet the mathematics means they can be understood in the same way.

For example, a liquid crystal placed under pressure may first alter its structure along a single straight line, and this is like a tear in the original structure. If we mentally excise this line defect – that is, create a virtual hole – the surrounding material then becomes *topologically* similar to a wedding ring. A loop around the line will not be able to contract to a point. In terms of Betti numbers, we can characterize this situation as $B_1 = 1$.

There are other configurations of topological defect too. It might be that all the molecules in a certain material are angled inwards towards a certain fixed location in the centre. This special position is known as a point defect, or a 'monopole'. By removing it, the surrounding space contains the same sort of hole as our beachball contains, and can thus be reflected through the second Betti number, $B_2 = 1$.

It is not merely that the vocabulary of topology can be brought to bear in these situations; understanding the topology behind these sorts of defects can bring real physical insights. For instance, line defects in a liquid crystal will not remain static, but flow around with the rest of the material. These may lead to highly complex scenarios in which two or more line defects become entangled. To understand when this will happen, and when two line defects will pass each other without becoming entangled, in 1977 two researchers, Valentin Poénaru and Gérard Toulouse, delved deeply into the underlying topology. Their analysis amply demonstrated the power of topology to settle difficult questions from physics. Similar ideas went on to illuminate other unusual states of matter, such as 'superfluids'. While ordinary fluids such as water or honey are viscous, meaning that they have internal frictional forces which slow down their movement, superfluids do not. The result is that they can flow straight through – or even climb over – obstacles which would stop any traditional liquid. Helium, which is a gas at room temperature becomes liquid at around −269°C. As it is cooled even further, towards absolute zero (−273°C), Helium becomes a superfluid. This process has been observed in the laboratory many times, but understanding how and why it happens has required an analysis of the topological defects which are likely to appear in the material.

The Big Bang and the early universe offer glimmerings of even more exotic types of topological defect. It is believed that phase transitions, taking place as the universe expanded from its original highly dense state, may have initially appeared as tiny bubbles, beginning as point defects, which evolved into a two-dimensional surface completely separating two phases of matter. As the bubbles grew and merged, the new phase gradually supplanted the old one.

These surface defects are known as domain walls. Excising these, as it were, completely divides the surrounding region into two parts, leading to a disconnected space. So, a single domain-wall defect can be captured by the zeroth Betti number: $B_0 = 2$. Under some models of physics, it is hypothesized that the early universe may have been divided into a patchwork of cells by a complex network of such domain walls.

Although this universal honeycomb has not been established for certain, domain walls *do* exist in a much more familiar setting: magnets. Within a magnetic material the magnetic field will vary from place to place, in one region pointing leftwards, and in a nearby region rightwards. These two regions may be divided by a narrow domain wall, across which the magnetic field twists sharply.

Stranger still, as topological defects go, are textures. While line defects are one-dimensional, and domain walls are two-dimensional, textures are theoretical three-dimensional defects. Textures could be captured by the third Betti number $B_3 \neq 0$ meaning that the underlying fields of physics curve around upon themselves along some higher dimension, possibly in some complicated knotted way. Physicists have theorized these as appearing when the universe evolved from its initial perfectly smooth, uniform state into one with the full variety of particles and forces that we know today, a process known as symmetry-breaking. Such textures have been proposed as ways of explaining fundamental mysteries of the cosmos. They include the nature of 'dark matter' – the invisible type of matter proposed as accounting for much of the mass in space – the rate at which the early universe expanded and the formation of galaxies. Exciting though the hypotheses are, analysis of the cosmic microwave background – the echoes of the Big Bang that we detect from all directions – has failed to show up any of the tell-tale hot and cold patches which a texture ought to leave behind.

All this cosmological conversation sounds a long way from experiments with a beachball and a bagel. Yet, whether or not the textures of theoretical physics ultimately play a major role in our future understanding of the universe, the remarkable thing is that, with the mathematics of holes now well developed, we are already equipped to investigate the concepts, and so know exactly what to look for. The mathematics is there for the taking, ready to be applied.

RAIN OR SHINE?

The mathematics of
weather forecasting

'Perpetually obsessed' would be an apt description of the British relationship with the weather. This may be an unfortunate stereotype, but there is some justification for the amount of analysis and conversation devoted to whether it will rain or shine. As islands lying where major weather systems battle it out for dominance, warmed by the waters of the Gulf Stream and subject to the fickleness of the Jet Stream high above, the UK is blessed – or burdened – with particularly changeable weather.

The challenge of understanding – and then predicting – the world's weather is huge, and our understanding still not perfect, as witnessed by the occasional mistakes in forecasting. Yet this subject taps into a rich mathematical field, which has contributed vital tools and methodologies not only to meteorology, but across a great deal of science. Galileo famously stated that the book of the universe is written in the language of mathematics; nowadays we can say, more specifically, that almost all of nature's laws are best expressed by one particular type of mathematical expression: a differential equation, which describes how systems change. Efforts to solve these equations have an illustrious intellectual heritage, and have provided the keys for numerous solid scientific predictions. But nowhere is this challenge greater or more important than in the attempt to grasp our planet's weather systems and climate. And it was a curious Englishman who began the whole business of mathematical weather forecasting.

Speeding up and getting high

Understanding how systems change, in general, has been a longstanding concern in mathematics, leading to several of its greatest triumphs. Yet today change remains a delicate and difficult topic, which requires not only mathematical and scientific insights but the intensive use of computer models to make predictions.

Evolving systems that are described by *difference equations* (see *The chaos in the fishpond*) are fascinating in their own right, as well as contributing valuably to disciplines such as population biology. For modelling physical systems, however, difference equations have their limitations. The trouble here is that they are fundamentally *discrete*, in the sense that they describe a sequence, such as 1, 1, 2, 3, 5, 8, etc., which begins with one number, then jumps to a second, and then a third and so on. The physical world, in contrast, is smoother; rather than hopping from one value to the next, physical processes tend to take in all the intermediate values along the way. In this continuous world, the role of *difference* equations is taken by *differential* equations, and innumerable physical processes – from a ball rolling down a hill to the laws governing our atmosphere – are best expressed using this language.

With the climate, as in science generally, the kind of change we are interested in may be over time – something speeding up from one moment to the next – or across space, such as the pressure in the atmosphere diminishing with height. The simplest example is movement. When we travel, our position changes, and a critical piece of information is the rate of change. If x stands for a cyclist's position, this rate of change is often written as x'. If x' is a small number then x is changing only slightly, and the bike is moving slowly. In fact, the value of x' is nothing other than the bike's *speed*. If the bike were travelling at a constant speed, then there is a familiar formula to calculate the answer:

$$\text{speed} = \frac{\text{distance}}{\text{time}}$$

or, rendered symbolically:

$$x' = \frac{x}{t}$$

where t is the time elapsed since the cyclist crossed the starting line.

This formula is fine for calculating the average (mean) speed over the course of a whole journey – so if the cyclist took 2 hours to travel 30 miles, then her average speed was 15 miles per hour. But this does not tell us her speed at any given moment, which itself is liable to change. To understand this, we need to know the rate of change of speed, in other words her acceleration, denoted as x''.

If we move from bicycles to air pressure in the atmosphere, a similar principle applies. The difference is that we now measure height instead of time. Unsurprisingly, the higher one reaches in the atmosphere, the lower the air pressure becomes. In fact, there is a subtle relationship between the pressure (p) at some height and the rate at which it changes there (p'), which is approximately expressed in the equation $p' = p \times 0.0001$.

Clearly, as we can see from the small number 0.0001, the rate of change of pressure p' is slow, so we do not notice any effect when we head upstairs in our homes. It is not always *imperceptibly* slow, of course. The pressure at ground level is approximately $p = 1{,}013$ hectoPascals (generally abbreviated as hPa, also known as millibars). From our equation, we know that the rate of change on the ground is approximately $p' = 1{,}013 \times 0.0001 \approx$ 0.1hPa per metre, enough for one's ears to 'pop' while driving quickly down a long hill.

The higher up one travels, the lower the value of p becomes. It follows that p' will become correspondingly smaller, and in turn p will reduce ever more slowly. Because this relationship is complicated, it is not obvious how to extract concrete predictions, such as the pressure at a particular height, say at 2,000 metres. To answer this sort of question requires an understanding of the science of rates of change – calculus.

The calculus revolution

Mathematicians have looked for ways of calculating x' from a knowledge of x since the time of Archimedes; but the major breakthrough had to wait until the 17th century and the independent work of two scientific luminaries: Isaac Newton and Gottfried Leibniz. Their new science of calculus was perhaps the single most important breakthrough in history when it comes to rendering mathematics useful to scientists in other disciplines.

What they found was a set of mathematical rules for determining x', starting with a mathematical expression for x.

For instance – returning to our cyclist – suppose that after one second she has travelled 1 metre, after two seconds 4 metres, and after three seconds 9 metres. The general rule here is that the distance is given by the time squared, symbolically $x = t \times t$ or $x = t^2$. But what would be her speed at any moment? Newton and Leibniz were able to answer this question, by way of the equation $x' = 2 \times t$. In other words after one second she is travelling at 2 metres per second (m/s), after two seconds 4 metres per second, and after three seconds 6 metres per second, and so on.

This is one instance of a more general rule, which says that if $x = t^n$ for some number n, then $x' = n \times t^{n-1}$. This rather non-obvious fact has been of immense value to scientists over the succeeding centuries. Applying the rule a second time tells us about the bicycle's acceleration: $x'' = 2$. In other words, this cyclist is speeding up at a constant rate of 2 metres per second *per second*.

The discovery of calculus immediately paid dividends in the physical sciences. Several of the most important insights also came from Newton, including his concept of the conservation of momentum. The momentum of an object is its mass multiplied by its velocity: a cycle weighing 10 kilograms travelling at 20 metres per second has a momentum of 200kg·m/s. Newton realized that in any closed system – one with no external forces acting on it – the total momentum of all constituent particles will remain the same, although it may be apportioned differently among them. A good example is snooker balls cracking into each other: if one hits another at 4 metres per second and then comes to a halt, conservation of momentum guarantees that the second ball must travel off at 4 metres per second, assuming that the two are the same mass. (If the second ball were twice as heavy – an unusual game of snooker – conservation of momentum tells us it that it would travel half as fast, at 2 metres per second.)

Such calculation must take into account the direction of travel. For example, if a 1-kilogram ball is travelling rightwards at 2 metres per second, and is hit head-on by another of the same mass travelling leftwards at 1 metre per second, the system's overall momentum is 1kg·m/s to the right. The conservation of momentum allows both balls to change direction in the collision, with the first heading back to the left at 1 metre per second and the second travelling rightwards at 2 metres per second.

As with many great breakthroughs, the discovery of calculus opened up as many new questions as it answered. The principle benefit that Newton and Leibniz had provided was a way to deduce information about the rate of change, x', from information about x. But in many scenarios, the need is the reverse – we have information about x' from which we need to find a direct description of x itself. This is where we get to the essence of differential equations – and we return to the thorny subject of the weather.

We met an example of the species above, the approximate relationship for the change in pressure: $p' = 0.0001p$. What a meteorologist would like is a simple expression telling us the pressure (p) at a given height (h), without having to worry about p' at all. This goal amounts to *solving* the differential equation. In this case, it is tractable. Again, Newton and Leibniz provided the tools for the job, which can be used to produce an exact answer, in the shape of $p = 1,013e^{-0.0001h}$. Here, e is the famous mathematical constant of around 2.718, and 1,013 features as the pressure at ground level (i.e. at $h = 0$).

Now if we want to know the pressure at $h = 2,000$ metres, we may plug this value into the equation to get an answer of:

$$p \approx 1013e^{-0.00012 \times 2000} = 1013e^{-0.24} \approx 1013 \times 0.8 \approx 800\text{hPa}$$

This is an approximate result – the equation is simplified and the figures rounded. Nevertheless the core message is that if we have the ability to solve the underlying differential equation, then we can extract real scientific information with little more than a pocket calculator.

As knowledge of the new subject of calculus grew, differential equations started to appear in ever more branches of science. A famous example came from the master 18th-century mathematician Leonhard Euler, who set himself a challenge. Newton's physics had been based on discrete, solid objects: two planets attracting each other gravitationally (see *The dynamic solar system*) or two snooker balls cracking into each other. Euler wanted to extend this to matter that was itself continuous. Instead of the particles of Newtonian physics, Euler wanted to understand the motion of fluids. This endeavour would found a new field of physical enquiry: fluid dynamics. And when applied to one particular fluid, the air circulating in our atmosphere, Euler's work would come to play a foundational role in the emerging science of meteorology.

Flowing fluids and troublesome turbulence

Euler's triumph was to construct a differential equation that described how Newton's principle of conservation of momentum translated to a flowing fluid. It was a subtle business, since the speed of flow might change across three dimensions (rightwards, forwards and upwards) as well as from moment to moment. The general idea was natural and intuitive: fluid will tend to flow from regions of high pressure into those of low pressure. It will also respond to external forces, of which the most significant is usually gravity. (Also relevant for the weather is the Coriolis Effect, the apparent force caused by the earth's rotation.) For Euler to derive the exact, correct expression required the science of calculus to be pushed to new extremes.

Euler's resulting Fluid Equation was a magnificent achievement, yet there was a notable omission: he had ignored the *viscosity* of the fluid, a measure of its thickness or stickiness. For highly viscous liquids like honey, this is clearly not a reasonable simplification; but neither is it for air, as anyone who has felt the effects of a strong gale will testify. It was in the 1840s that this omission was put right by the French engineer Claude-Louis Navier and an English physicist, George Stokes, who separately tackled the problem. They incorporated an additional term into Euler's equation to account for a fluid's viscosity. Their findings, encapsulated by the Navier–Stokes Equations, have played the central role in our analysis of fluids ever since.

There is a deep paradox here, however. From a mathematician's perspective, the resulting equations are so difficult as to be literally insoluble, at least to date: it has been more than 150 years since Navier and Stokes's work, and in all that time no-one has been able to find a single solution to the equations – which is to say that no-one has been able to produce a mathematical description of a fluid that obeys the Navier–Stokes laws (leaving aside trivial solutions, such as a completely stationary fluid). Indeed, this remains one the Clay Institute's Millennium Prize conundrums, announced in 2000, with a bounty of $1 million on offer for whoever can solve it first.

The Navier–Stokes story illustrates a recurring theme in the relationship between science and mathematics. First, a combination of careful thought and painstaking physical observation suggests a new scientific principle; then, when translated into mathematics, this law typically takes the form of a differential equation. The finale should ideally be for scientists to solve the resulting equation, allowing humanity to exploit the discoveries and cash in with robust predictions – and this is where things can become seriously problematic. Indeed, many equations describe systems that are fundamentally chaotic.

In the context of fluids, the obstacle comes in the shape of turbulence, the phenomenon of fluid moving in complex, choppy ways, instead of smoothly, as occurs in rough seas or in winds near the surface of the Earth as opposed to the relatively tranquil upper atmosphere. This behaviour, which is very common, is extremely difficult to predict. Slow-moving fluid is not turbulent, because of the internal viscous forces dampening it down. But beyond a certain speed threshold, turbulence becomes nearly inevitable, a progression summed up in the ditty of pioneering meteorologist Lewis Fry Richardson:

Big whorls have little whorls
That feed on their velocity,
And little whorls have lesser whorls
And so on to viscosity.

It is this chaotic nature of turbulent air-flow that is encapsulated in Edward Lorenz's famous butterfly effect: the idea that a butterfly flapping its wings in Brazil can, in time, lead to tornadoes in Texas. This is what today's meteorologists are up against, with turbulence described by the eminent quantum physicist Richard Feynman as 'the most important unsolved problem of classical physics'.

The foundations of forecasting

This is not to say, of course, that science – or meteorology specifically – has hit the buffers. After all, today's weather forecasters successfully make increasingly detailed and accurate predictions. Physical observations may have given rise to equations that are, as yet, beyond our powers to solve, but it is at this point that the disciplines of mathematics and science diverge. While mathematicians continue to probe the Navier–Stokes Equations for exact answers, other branches of science are satisfied by *approximate* solutions, employing – with the intensive use of computers – an array of methods known as numerical analysis. These techniques are in constant use across a range of modern sciences and industries, but nowhere more so than in weather forecasting.

People have been second-guessing the weather for millennia, but the first truly mathematical forecast was carried out by Lewis Fry Richardson between 1916 and 1918. As a Quaker and a pacifist, during the First World War Richardson went to work on the Western Front as an ambulance driver. In this unlikely and dangerous setting, he not only acquired a reputation for diligence under fire, but also began serious research into mathematical meteorology.

Richardson's aim was to predict changes in the weather at two specific locations. To do so, he had to know the initial conditions in the surrounding regions. Luckily, weather monitoring had made great advances in the previous few years, and Richardson had access to a table of detailed readings recently compiled by another pioneer, Vilhelm Bjerknes. Crucially, this data was not limited to the surface of the earth, but included readings at various altitudes from 193 weather balloons at locations across Europe.

To analyse how air can change from moment to moment, Richardson amassed a battery of seven numbers to describe the air at any given point. First there is the momentum of the air (i.e. the wind) in three dimensions: north, east and upwards. Then there are the further four factors of air pressure, humidity, temperature and density. Tying these together is an intricate system of differential equations, describing how these elements affect one another and quantifying how each varies with time and across the three dimensions of space. Central are the Navier–Stokes Equations, but there are others too, emanating from the science of thermodynamics (see *Hot stuff*), which describes how the energy absorbed from sunlight is then redistributed around the atmosphere, causing heat sources and sinks. The continuity equation meanwhile states that the total amount of air in the atmosphere is unchanged – none is created or disappears.

Additionally, Richardson adopted the ideal gas law, which relates the pressure (p), temperature (T) and density of air (d) at a single point. Specifically, it says that, although these three numbers may vary from moment to moment, it will always be true that $p = 287 \times T \times d$, the number 287 being the 'gas constant' for dry air (for other gases, it varies).

Altogether, this armoury amounted to a somewhat terrifying collection of equations. But Richardson added one other, which plays a simplifying role. The reason that we can breathe on the surface of the earth is that air is held in place by gravity, rather than dissipating into space. It does not all sink to ground level, because counterbalancing gravity is the air pressure. The hydrostatic approximation is the assumption that these two cancel each other out exactly. In equation form it says $p' = -g \times d$, where p' is the change in pressure at a certain height, g is the force of gravity and d is the air density at that point.

Finally, in order to make his forecast, Richardson now had the unenviable task of attempting to solve the resulting system of seven equations, with all of Bjerknes's data incorporated. And he had to do it without the modern meteorologist's best friend, the computer. He followed the standard approach of numerical analysis, which in essence is the replacement of continuous differential equations with a discrete system. So Richardson divided a map of Europe into a grid, and analysed the conditions at the centre of each square. Instead of dealing with every instant of time, he worked in steps of three-quarters of an hour. From the initial data, the equations told Richardson the rates of change of all the factors involved.

The birth of weather forecasting

In terms of *methodology*, Richardson's work was a triumph: not only did he assemble the correct system of equations, but he was able to come up with a workable numerical method to find approximate solutions to them. As a *practical forecast*, however, it was a disaster. To start with, the lengthy series of complex

PREDICTING CHANGE: THE METHOD OF NUMERICAL ANALYSIS

To illustrate the basic principles of numerical analysis, we do not need to tussle with Richardson's atmospheric equations – we can take a simpler route and call in our cyclist. Suppose it is reported that her speed in metres per second is exactly half her distance (in metres) from the start of the road at every single moment. Expressed mathematically, we get

$$x' = \frac{1}{2}x$$

This is our fundamental equation. We might also know that at the moment we start the clock she is 1 metre down the road. That is to say, when $t = 0$, $x = 1$. So the equation tells us that at this moment she is travelling at half a mile per second, as given by

$$x' = \frac{1}{2} \times 1 = \frac{1}{2}$$

The next step is to ignore our original equation, and instead extrapolate this rate of change into the future, to provide the next value. If our cyclist continues at this speed, then one second later, when $t = 1$, she will have travelled a further $\frac{1}{2}$ metre. So, the approximation here is that when $t = 1$, then

$$x \approx 1\frac{1}{2}$$

We can then feed this value back into the equation, to find the next rate of change:

$$x' \approx \frac{1}{2} \times 1\frac{1}{2} = \frac{3}{4}$$

And then extrapolate *this* speed into the future for another second, and so on. This process can be repeated as many times as necessary, and so a sequence of numerical predictions emerges.

calculations took him at least six weeks to complete, for a forecast of just six hours into the future! What is more, his figures indicated a rise in pressure of 145hPa over the course of those six hours, a wildly unrealistic figure, equivalent to descending a metre and a half under water. It is a testament to his confidence as a scientist that, despite these major shortcomings, Richardson had the confidence to publish his work. It constituted, in the words of Peter Lynch's 2006 book *The Emergence of Numerical Weather Prediction*, 'one of the most remarkable and prodigious feats of calculation ever accomplished'.

Decades later, the large ENIAC machine (see *All-conquering algorithms*) would be put to use to prepare the first computerized weather forecasts. Over the years, this numerical approach to forecasting has revealed many familiar meteorological features, such as cyclones (the regions of low pressure around which winds circle) and anticyclones, the high-pressure equivalents. It has been discovered that, because of the Coriolis Effect, cyclones typically rotate anticlockwise in the Northern Hemisphere and clockwise in the Southern, while anticyclones do the opposite.

Nowadays weather satellites provide us with masses of detailed data, while meteorologists employ the world's most powerful supercomputers, along with the latest advances in numerical analysis, to solve the system of atmospheric equations. We all know prediction is not a perfect art; yet the increasing sophistication and accuracy of results is a testament to the considerable progress made. And the essential techniques used in making today's forecasts began, appropriately enough, with a weather-obsessed Englishman, in the shape of Lewis Fry Richardson.

SMOKE AND MIRRORS

Statistical illusions and numerical mirages

Benjamin Disraeli, the flamboyant 19th-century prime minister, may or may not have been the originator of the famous assertion that 'There are three kinds of lies: lies, damned lies, and statistics.' Whoever coined the phrase, there is considerable truth in the view that statistical data is liable to misuse and misconstruction.

At the same time modern society is hugely information-rich. It is constantly gathering data about us, from companies gauging the demographic appeal of their products, to national censuses describing the social make-up of a country, to sports' analysts calculating players' averages. Analysed properly, these mountains of data certainly can yield valuable results. But they pose risks too, whether through accidental misinterpretation or wilful misrepresentation. Sometimes an eye-catching new social trend may be nothing more than a statistical illusion.

Many of these bogus statistics are surprisingly durable, because 'bad' statistics can exude a kind of folk conviction, seemingly according with common sense. It is only by bringing mathematical rigour to bear that we can hope to spot the flaws in the arguments.

Sins of omission, sins of commission

1914 is a useful place to start. At the beginning of the First World War, British and French soldiers were sent to the trenches with no head protection, only cloth caps. A year later, once manufacturers had had the chance to catch up, steel helmets were issued. But this seemingly sensible move produced a profoundly unexpected consequence. Field hospitals suddenly recorded a dramatic *increase* in the number of head injuries. How could it be that helmets were having exactly the opposite effect from the one intended? The answer, of course, is that they were not. To understand why, it is enough simply to take note of what was being counted: *injuries*, not fatalities. With the introduction of the helmet, soldiers who would have otherwise died from their head-wounds were surviving beyond the battlefield.

This is a good example of the principle that *correlation does not imply causation*. Technically, correlation is a statistical relationship between two phenomena – in this case, the introduction of helmets and the increase in head injuries. To take a more pleasant example, sales of sun-lotion are *positively* correlated with sales of ice-cream – 'positive' here meaning that an increase in one term will mean a likely increase in the other. (A *negative* correlation matches an increase in one term with a decrease in another.) It is not, of course, likely that the sensation of sun-lotion on the skin somehow encourages a desire for ice-cream, or that the taste of ice-cream gives a psychological cue to go out and buy sun-lotion. Rather, the sales of ice-cream and sun-lotion are each independently triggered by an additional, third factor: sunny weather.

Although the distinction between correlation and causation

seems simple enough, it can lead to serious statistical abuse, whether accidental or wilful. All it requires is for data to be presented in a way that omits some central factor, as sometimes happens with over-exuberant advertising. A claim by, say, a breakfast-cereal manufacturer that 'consumers of Fuzzy Flakes have lower cholesterol than the national average' may on the face of it sound impressive. But perhaps Fuzzy Flakes come in a brightly coloured box, decorated with pictures of a famous children's TV character – and thus the cereal is targeted at young children, who, by virtue of their age, have naturally lower choles-terol than the national average. Or it may even be that Fuzzy Flakes come in a box laden with photographs of fit adults jogging in the open air, and is targeted at health-conscious. They, too, are likely to have lower than average cholesterol. In both cases, the manufacturers are able to piggy-back onto a low-cholesterol demographic, thus allowing them to make a claim that, though not actually untrue makes an unjustified insinuation about the health benefits of their product.

Advertising is, in some ways, an easy target – indeed, a serial abuser. But at least lives are not usually at stake. In other, and more serious, contexts it sometimes requires lateral thinking to extract the correct message from a particular scenario, because the obvious conclusion may be the wrong one. A famous example occurred during the Second World War with Abraham Wald, a statistician who analysed the damage sustained by US bombers. A study had been performed, analysing the number of bullet holes in different regions of the returning aircraft. Across the fleet, patterns emerged: some regions of the planes, such as the main fuselages, came back peppered with far more hits than others, such as the engines. This prompted a proposal to strengthen what appeared to be the more vulnerable, more bullet-riddled, areas. When Wald was deployed to examine the problem, he saw that this conclusion had it entirely wrong. He suggested instead that badly shot regions (including the fuselage) should be left as they were, and the regions showing little or no damage, such as the engines, should be reinforced.

It was a brilliant insight, and one that would save the lives of many service personnel during the war and in the years to come. Wald realized that because investigations were limited to the bombers that had safely made it home – as opposed to those that had not – the real question ought to be: what patterns of damage would the *downed* planes reveal? Although a direct investigation was impossible, a little statistical argument provided the answer: they would surely show bullet holes in the engines and other regions under-represented in the surviving planes. So damage to the fuselage might be withstood, as the numbers of surviving planes showed, but damage to the engines could be fatal.

Statistical phantoms

Sometimes the real challenge is to search though different layers of data to find the truly meaningful numbers. In 1973, the University of California at Berkeley was sued for alleged bias against women in its admission process for postgraduate students. At first sight, the evidence seemed compelling. Of the candidates who made it to the final round of selection, 44 per cent of male applicants were successful compared to only 35 per cent of women. The disparity looked bad for the university; at the very least, it warranted further investigation.

When the statistician Peter Bickel and colleagues examined the data in the following year, they discovered something very unexpected indeed. Their first step was to break the data down into different university departments, looking for the source of the bias. But as they did so, the bias itself mysteriously evaporated. In their investigations of the decision boards in the various departments, they found 'few decision-making units that show statistically significant departures from expected frequencies of female admissions, and about as many units appear to favor women as to favor men'. In fact, the overall the picture was exactly the opposite of what they first expected – 'a small but statistically significant bias in favor of women'. The university was exonerated, yet the mystery remained, for the original data was correct.

The answer lay partly in the old mantra that 'correlation does not imply causation'. But here also was a striking example of the phenomenon known as the Yule–Simpson Paradox, named after two British statisticians: Udny Yule, who wrote about it in 1903, and Edward Simpson, who elaborated the paradox in 1951. As with our other examples, the misleading picture arose from a failure to take into account an important aspect of the scenario.

Bickel and his colleagues used the metaphor of fishing. Suppose two nets are lowered into a river: one has a coarse mesh while the other is finer. Needless to say, the finer net will catch more fish – say 60 per cent of those that swim through – while the coarser net may trap only 25 per cent. If the river is populated by male and female fish of a particular species, and if we assume that there is no difference in size between the sexes, then there can be no built-in bias towards either sex. However, there may nevertheless be an overall imbalance stemming from an additional factor.

Let's imagine that female fish have a tendency to swim down the centre of the stream, while male fish travel closer to the riverbank. If the *coarse* net is dropped in the centre of the stream for an hour, more female fish than males will swim here: perhaps 80 females and 16 males. The *fine* net is dangled near the bank for the same period of time, where 20 females and 100 males swim by. The result is that more male fish will be caught overall, even though neither net has any inherent bias towards either gender. The picture might look like this:

	Female fish caught	Male fish caught
Coarse net, near centre	$\frac{20}{80}$ (25%)	$\frac{4}{16}$ (25%)
Fine net, near bank	$\frac{12}{20}$ (60%)	$\frac{60}{100}$ (60%)
Totals	$\frac{32}{100}$ (32%)	$\frac{64}{116}$ (around 55%)

In other words, from two perfectly fair nets, we have a result heavily imbalanced towards catching male fish. Even if we alter the figures, to presume that both nets have an inherent tendency towards catching females (perhaps because they are slightly larger than males), nevertheless the net-placement factor can easily outweigh this:

	Female fish caught	Male fish caught
Coarse net, near centre	$\frac{30}{80}$ (37.5%)	$\frac{4}{16}$ (25%)
Fine net, near bank	$\frac{15}{20}$ (75%)	$\frac{60}{100}$ (60%)
Totals	$\frac{45}{100}$ (45%)	$\frac{64}{116}$ ($\approx$55%)

The Yule–Simpson Paradox is visible here: both nets have a tendency to catch more female fish than males, but this trend is reversed when the figures for the two nets are combined. In the case of the Berkeley's graduate schools, the additional factor analogous to the placement of fishing nets was the difference in the types of department that attracted male and female applicants. For whatever reason, women were applying in larger numbers to departments that were more highly oversubscribed, and therefore which rejected a greater proportion of applicants; by contrast, more men were applying to the less competitive departments. (It is an interesting aside that the authors identified mathematics as a major factor here: 'The graduate departments that are easier to enter tend to be those that require more mathematics in the undergraduate preparatory curriculum.')

Like the Yule–Simpson paradox, 'regression to the mean' can explain statistical mirages, deflating unsupportable claims. It also reminds us of a hugely important, but often overlooked, part of life: luck.

A bit of luck is perennially useful for anyone sitting an exam, so let us imagine two girls, Annabel and Betty, doing exactly that. The national average score, we are told, is 50 per cent for the two exams they both sit. In the first exam, Annabel does very well, scoring 88 per cent. In the second she is less successful, but still scores above average, at 70 per cent. Betty meanwhile struggles badly in the first exam, scoring just 25 per cent. In the second she does significantly better, but still below average, earning 43 per cent.

It is easy to imagine the sorts of interpretations parents or friends might invoke to explain these results: Annabel became overconfident and complacent after her initial triumph, and didn't try as hard in the second. Betty got a terrible shock in her first exam, and so applied herself more seriously in the second. It may be that these assertions contain elements of truth; what they miss, however, is that this sort of phenomenon is to be expected *for purely statistical reasons*. How well someone performs in an exam – or in a football match, business venture, or romantic relationship; anything, really – will depend on some combination of ability, effort and preparation, but also luck.

Let's imagine for a moment that the girls' exams are not really exams at all, but products of purely random chance: there are 100 multiple-choice questions, and in each case each girl has a probability of 50 per cent of picking the right answer from two options. If we had to guess the two girls' scores in advance, our best bet would be 50 per cent. But Annabel gets lucky, and 'does well' in the first exam, achieving 88 per cent. What if we then predict her score in the second exam? Well, nothing has changed, statistically speaking, and the best guess remains 50 per cent. From this perspective, it is the *very fact* that Annabel got a high score first time around that makes her score likely to drop in the next exam. Similarly, Betty's low first score of 25 per cent is highly likely to be exceeded by her second. In both cases, the chances are that the girls will move from their extreme positions towards the mean.

The same principle holds even when the exam scores are not pure luck, but – as would normally be the case – some combination of luck, ability and other factors. Let's suppose (somewhat artificially) that of the 100 marks available on the exam, sheer knowledge accounts for 70 and luck for 30. We can assume the girls' knowledge to be static, say with Annabel on 60 and Betty on 20.

We can see here that, in the first exam, Annabel was very lucky (achieving 28 out of 30) and Betty very unlucky (gaining only 5 out of 30). So it was to be expected that they would both move towards the mean in the second exam.

	Annabel			Betty		
	Knowledge /70	Luck /30	Total /100	Knowledge /70	Luck /30	Total /100
Exam 1 results	60	28	88	20	5	25
Exam 2 results	60	10	70	20	23	43

Regression to the mean is a widespread, but hugely underappreciated phenomenon. It occurs in many guises, specifically whenever two phenomena are correlated but not perfectly so. It is particularly common in sport, though rarely picked up by commentators. For example, a long-jumper who does an extremely good first jump is, merely by virtue of this fact, likely to perform less well second time around. In 1989, the magazine *Sports Illustrated* noted that 90 per cent of the baseball players who hit over 20 home runs in the first half of the season failed to repeat the feat in the second half. Statistically, it is unremarkable.

That same magazine provides another notorious example of regression to the mean – one frequently replicated in the cases of other beneficiaries of unusual success. A sportsperson who has an exceptionally good week may find themselves on the magazine's cover. The following week, it has often been noted, they are likely to suffer a decline in form. Is this some sort of supernatural jinx visited on those who adorn *Sports Illustrated*? No. Having over-achieved, they are just moving back to the natural order of things.

After all, correlation does not imply causation.

WHERE ARE WE?

GPS geometry and
Einstein's explorations

In order to operate one of London's famous black taxi cabs, a would-be licensed cabbie needs to acquire The Knowledge – an encyclopaedic understanding of London's streets, which takes up to four fearsomely difficult years to acquire. For the public wanting a ride, there has always been the option of taking a minicab, whose drivers are Knowledge-free. This requires you to run the risk that your driver might have no clue how to get to your destination – indeed may never have heard of it.

But now there is sat-nav. Thanks to the Global Positioning System (GPS), a network of 24 satellites (plus 3 spares) launched by the US Defence Department and operational since 1994, arguably anyone can, at the very least, know where they are; and minicab drivers equipped with satellite-navigation aids can simply type in the details of where they want to go, and follow the instructions given.

Luckily for black-cab drivers, stories of sat-nav misuse and incompetence abound, and occasionally are even newsworthy; and the technology cannot yet provide the clever shortcuts around traffic blackspots that cabbies know from experience

– not to mention the colourful conversation for which they are famed. Nevertheless, these stories entertain because they are the exception to the rule. GPS has transformed our ability to orient ourselves in our environment, and it should be no surprise that this world-changing technology is heavily indebted to mathematics. Indeed, the basis for the whole concept is provided by a few essential geometrical ideas, along with a sprinkling of insights from the science of relativity.

Compasses in space

The GPS system relies on a physical property of our universe: that radio waves travel at a fixed speed, specifically 299,792,458 metres per second, known as the speed of light or c for short. This fundamental fact allows distances to be gauged, so long as we are able to measure time precisely enough. For instance, suppose we want to calculate the distance between a radio transmitter and a radio receiver. One way to do this might be to send a message from the transmitter, the only content of which is the exact time it was sent. The receiver can then compare that message to the clock-reading at the moment the message arrives, with simple subtraction to reveal the duration of the journey. If we call this time t, then typically, because light travels so fast, t will be a very small number – unless the receiver and transmitter are situated on different planets. Let us imagine $t = 0.00015$ seconds.

We also know the speed at which the message travelled, since the answer is always c. It is now a simple matter to calculate the distance the message has come, using the old standby formula: distance = speed × time. In this case, the distance will be

$c \times t = 299{,}792{,}458 \times 0.00015$, which is around 44,969 metres, or just under 45 kilometres.

This calculation gives the central idea of GPS. The system's network of 24 satellites orbit the earth at an altitude of around 20,200 kilometres in such a way that, at any moment, every point of the earth is within sight of at least 6 of them. Each satellite constantly broadcasts the exact time and its precise location, so that receivers on the surface of the planet can perform the same sort of calculation as we have just done.

However, an extra consideration is needed to pinpoint our position. It is not enough simply to know that we are 45 kilometres from a particular satellite (say S), since there are various places which satisfy that same description. In fact, the collection of all positions 45 kilometres from S will form a circle on the earth: we could be anywhere along that circle. If we add in the possibility that we are in an aeroplane or spacecraft, then we could be anywhere on the sphere of radius 45 kilometres, centred on S.

This problem is solved by the process known as trilateration, which uses the information from several satellites, and can be illustrated by a further example.

Suppose someone picks a point on a map and challenges us to find it from a few snippets of information. The first piece of data we are given is that the position is 5 centimetres from a fixed point A (representing a beacon) – so we know it must be somewhere on the circle with radius 5 centimetres, centred at A. If we are additionally told that it is 10 centimetres from a second beacon B, we will look at the places where the new circle (radius 10cm, centred at B) meets the first. Typically, there will be two such points; occasionally there will be one, if A and B are exactly

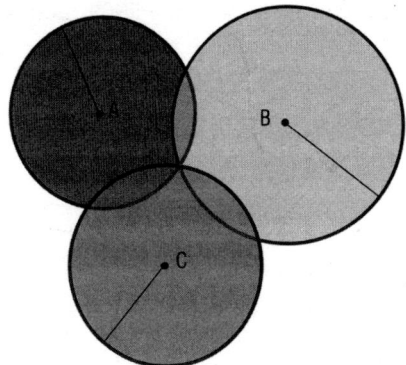

The process of trilateration locates a position from knowledge of its distance from three beacons.

15 centimetres apart – in which case we have already found the map location. (If the two circles miss each other, we have been set an impossible task!)

Having narrowed things down to two possible points, just one further piece of information should be enough to decide which is correct. If we are told the location is also 8 centimetres from beacon C, then – we hope – only one of the two points will satisfy that condition. It can only happen that *both* points satisfy the final requirement if A, B and C are arranged in a perfectly straight line, but this would constitute a serious design flaw in the beacon network!

Exactly the same idea works in three dimensions, and this forms the basis of GPS. For it to work, four satellites are needed: the first reading places our location on a sphere, centred at one satellite; the second narrows this down to a circle on the surface of that sphere, where it meets a second sphere; introducing a third sphere will again pin it down to one of two points (since the satellites will never lie in a straight line); and the fourth identifies it exactly.

The ancestors of GPS

GPS did not spring out of nowhere, of course. Its forebear was a system of beacons used to help ships at sea ascertain their positions. LORAN (standing for LOng RAnge Navigation) was devised and built in the United States during the Second World War, and it continues to be used in some parts of the world today. The principle is similar to GPS trilateration: each beacon contains an atomic clock, and a ship will tune into three of them in order to deduce its location. The difference, however, is that rather than broadcasting the time, as GPS satellites do, they simply emit a pulse, each on their own unique frequency, at a regular interval. Of course, such a pulse does not carry enough information for the ship to calculate the distance to the beacon directly, but the ship can compare the times at which the pulses from two beacons arrive.

To simplify somewhat, we can imagine that two beacons A and B emit pulses simultaneously. These will arrive at the ship at different times: say a and b. The relevant information is the gap between these arrival times, so mathematically it is $a - b$. Although the ship cannot calculate its distance from either of the beacons, it can use this information to calculate the *difference* in these two distances. If this gap is $a - b = 0.00015$ seconds, then it might deduce, for instance, that it is around 45 kilometres further from A than it is from B.

When plotted on a sea chart the possible locations that tally with this data do not form a circle but another curve known as a hyperbola. Another pair of beacons, perhaps B and C, will produce a second hyperbola, and the ship's location will be where these two curves cross.

LORAN's principle can also be used in reverse, with detectors at fixed points and the emitter in the unknown location. Such audio-location systems were used as far back as the First World War. In order to discover a hidden enemy gun position, three audio sensors were set up, attached to a central clock, which compared the time at which the bang reached them. The time-differences between these readings placed the gun at a position on two intersecting hyperbolas, giving its location. This was a major element of military technology up until the adoption of radar (RAdio Detection and Ranging) during the Second World War. Today, sonar (Sound Navigation and Ranging) remains in use underwater where sound waves travel further and faster than they do in the air.

For today's GPS system to become a reality, various technological breakthroughs were needed – most obviously, satellites. But also, both LORAN and GPS rely on highly accurate clocks. Without these, neither system stands a chance. (Audio-location systems could get away with less accuracy, since they work with slow-moving sound waves rather than the speed of light.) So, LORAN beacons and GPS satellites both carry atomic clocks – devices that measure time by the number of fluctuations of electrons orbiting caesium atoms, the first functioning example of which was built by Louis Essen in 1955.

Atomic clocks are evidently impractical for receivers in mobile phones or car-navigation systems, so here ordinary quartz clocks are used, which are less accurate. Luckily, minor discrepancies can be overcome by the observation that if the clock in the receiver is slightly wrong, it is equally wrong with respect to *each* of the satellites. So long as the magnitude of the error is not too large this can be spotted (and automatically corrected) without too much difficulty.

Enter Einstein

A potentially more troublesome source of sat-nav error brings us to Albert Einstein's theory of relativity. People are often surprised that Einstein's abstruse ideas have practical, observable consequences for domestic gadgets; but the plain fact is that if relativity were not factored in, the GPS system would quickly become largely useless.

There are two relativistic effects in GPS, stemming from the two branches of the subject: special and general relativity. The starting point of special relativity is that being stationary – 'stationarity' – is not fixed or absolute, but is rather a relative concept: to an observer looking at a railway, a passing train seems to be travelling quickly, while she believes herself stationary, standing on a static bridge. A passenger sitting comfortably on the train, however, also feels stationary, while the bridge and woman waving on it appear to fly past. It has been known since the era of Galileo that this is no illusion: anyone who is moving at constant speed is indeed stationary according to their *own* reference frame, and is in motion according to any other. Through the work of several physicists, including Albert Einstein and Hendrik Lorentz, the early 20th century brought a deeper understanding of how measurements of distance, time and mass differ when taken in respect of different reference frames. The discoveries made here were startling when first unveiled, but have subsequently been supported by several decades of experiment.

For GPS, the most important aspect of the theory is how two clocks will appear to each other, if one is moving relative to the other. The surprise is that relative to a 'stationary' observer, a clock moving at a high speed will appear to be running more slowly than a stationary clock. This is exactly the situation regarding the atomic clocks on satellites.

The primary technical tool for translating between two reference frames is a Lorentz transformation. It asserts that if the satellite is travelling with a speed of v, then its clock will seem slower to a stationary observer on the Earth by a factor of a, where

$$a = \frac{1}{\sqrt{1 - \frac{v^2}{c^2}}}$$

Here, c represents the speed of light as usual, and v is the speed of a GPS satellite, which is around 3,900 metres per second. Putting these numbers into the formula, a comes out at around 1.00000000008. This means that over the course of one day (or $24 \times 60 \times 60 = 86{,}400$ seconds) the satellite's clock will lose around $0.00000000008 \times 86400 \approx 0.000007$ seconds, or 7 microseconds.

While special relativity concerns objects moving relative to one other, general relativity is the science of gravity. One of its predictions is that a clock near a heavy object, such as the earth, will seem to tick more slowly than one further away. (It is worthwhile noting that, with special relativity the argument would be equally valid if we mentally swapped which of the clocks was stationary and which moving; but, unlike motion, gravity is not a relative phenomenon, but an absolute.)

Here, the role of the Lorentz transformation is played by the

Schwarzschild metric, named after Karl Schwarzschild, who, in 1915, studied Einstein's recent work to extract several concrete predictions. Many years after his death, Schwarzschild's work would become highly relevant to the study of black holes.

The essential fact here is that, compared with a gravity-free region, a clock near a heavy body (such as the earth) will seem to tick more slowly by a factor of b, where

$$b = \sqrt{1 - \frac{2GM}{rc^2}}$$

In this formula, G is the fixed number called the 'universal gravitational constant' (around 6.7×10^{-11}), while M is the mass of the earth (approximately 6.0×10^{24} kg), and r is the distance from the clock to the centre of the earth. (For more on gravity, see *The dynamic solar system*.)

In the case of GPS it is necessary to apply this formula twice: once to the receiver on the ground, and once to the orbiting satellite. Crucially, r will differ between these two. On the earth's surface, r is approximately 6,371,000 metres. Meanwhile, satellites orbit at around 20,184,000 metres above that, making r about 26,555,000 metres here. These two values of r produce different values of b for the receiver and for the satellite, and we need to subtract one from the other to find the difference. Plugging in all the numbers, the result is a disparity of around 45 microseconds (0.0000045 seconds) per day.

While grappling with relativity is an unavoidable challenge for GPS, it does not need to enter the fray for navigational beacons. Since both the beacon and the ship are on the earth's surface (and therefore the same distance from its centre), general relativity is irrelevant; and the speed of a ship or boat is so slow as to make special relativity not worth bothering with. For GPS however, relativistic considerations are absolutely crucial. In fact these two corrections, special and general, work in opposite directions. So the overall error to be corrected for is around $45 - 7 = 38$ microseconds per day.

With this correction made, GPS becomes the reliable and ubiquitous system depended upon by people around the world. It not only helps tourists lost in foreign cities, but comes into play whenever positioning and timing are critical: the automatic detection of time zones in our mobile phones; the monitoring of tectonic plates and shifts during earthquakes; the hunt for survivors in the aftermath of disasters; tracking aircraft through the skies; the tagging of property, pets and prisoners; a host of military applications, from behind-the-lines insertion to search-and-rescue operations – and, of course, for helping minicab drivers become better at their jobs.

MORE BANG FOR YOUR BUCK

Optimizing the world

'There is no such thing as a free lunch.' So the saying goes, reminding us that everything comes at a cost. Of course, the search for free lunches is an abiding human trait. A small industry of gurus and self-help books feeds this appetite, but the truth remains that most things worth having are worth paying for, either with money, or through an investment of time and effort. Everyone who has managed a business or handled a large project knows that trade-offs have to be made, and the more that there is at stake, the more complicated things become. The question is not whether a particular trade-off is worth making, but rather: what is the best trade-off to make in this situation?

Mathematics plays a crucial part in answering this question. The dilemmas we face often involve juggling large numbers of different factors: time, people, money and other resources. In searching for the optimal blend in a particular instance, we are often trying to *maximize* something – the output of a factory, the number of diners served in a restaurant, the availability of nurses in a hospital. In other scenarios, we may be propelled by a need to *minimize* something – a business traveller's carbon footprint, the costs of defending the nation, a school meals budget.

A million other examples could be cited. What they have in common is that mathematicians see them as optimization problems, the study of which has been one of the major practical applications of mathematics over the last century. The reason that they are not straightforward is that these problems come with constraints attached, boundaries imposed by time or resource limitations, or by essential requirements that must be met. Clearly, minimizing the school canteen budget by serving just chips daily or maximizing the guests in a restaurant by removing tables and chairs and making everyone stand to eat will both fundamentally fail.

So, the challenge is to optimize one quantity, subject to the relevant constraints. Developed in the 1940s, notably by Leonid Kantorovich and George Dantzig, the most popular approach to these types of problem is known as 'linear programming'. This term dates from before 'programming' became synonymous with computer science, and in this context carries the sense of a 'programme' as in a schedule or timetable. However, optimization studies grew side-by-side with the rise of the programmable computer, and together they play a pivotal role in humanity's ability to solve complex logistical problems.

Optimizing your aardvark

We can illustrate the general idea by way of an example in manufacturing – say a toy factory that produces two types of soft toy: aardvarks, for the adventurous child, and for more traditional customers, bears. If we imagine they create A hundred aardvarks and B hundred bears per day, the company needs to know what values of A and B to aim for. The first important thing to gauge, for optimization purposes, is how much the two toys are worth. If they sell for the same price, the factory should aim to maximize the number $A + B$, the total number of toys created per day (in hundreds): a central quantity called the target function. (If, though, an aardvark sold for twice the price of a bear, the appropriate target function would be $2A + B$.)

The resources needed to construct each of the toys are also relevant. We can assume, for simplicity, that each toy is constructed in two steps. First, it needs to be stitched on the sewing machine, and then filled on the stuffing machine. Because the toys are different shapes, these tasks require different lengths of time: let's say that 100 aardvarks require 1 hour for stitching and 2 hours for stuffing, while 100 bears require 3 hours for stitching and 1 hour for stuffing. And now we need to factor in the availability of the machinery, say 9 hours per day for the stitching machine, but 8 hours – because it takes longer to set up and refill – for the mechanical stuffer.

Setting this out mathematically, the total stitching time needed for A hundred aardvarks and B hundred bears is $A + 3 \times B$. Since there are at most nine hours available, then this quantity cannot exceed nine, which provides us with the first constraint: $A + 3B \leqslant 9$. We get a similar expression from the stuffing machine, giving our second constraint: $2A + B \leqslant 8$. Beyond these, there are two fairly obvious constraints to recognize, namely that the minimum that either A or B can be is zero, which can be expressed as: $A \geqslant 0$ and $B \geqslant 0$.

At this point we are well placed to answer a basic question: what are the *possible* values of A and B, subject to these

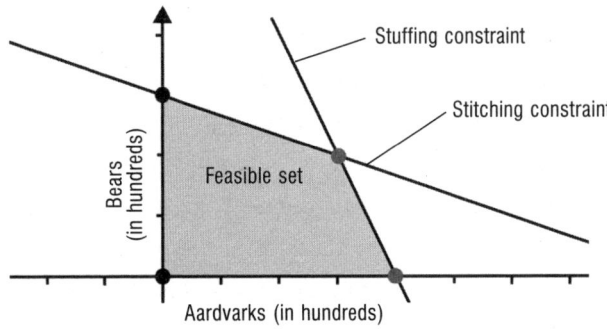

The feasible set of the toy factory, as expressed geometrically.

constraints? This prompts the observation that gives the whole subject its flavour, because the answer is best given *geometrically*. The possible values of A can be positioned along a horizontal axis, and those of B along the vertical axis, giving us a graph on which the coordinates of every point on the plane represent a pair of values (A, B). But of course, not all pairs will be compatible with the constraints we have identified, so the question is: which ones *will* be?

To begin with, the trivial constraints – the minimum production of zero toys, that is $A \geqslant 0$ and $B \geqslant 0$ – mean we only need to worry about the top-right quadrant of the graph, ignoring anything to the left of the vertical axis (where A is negative) or below the horizontal axis (where B is negative). The other constraints are a little subtler. The threshold for the stitching machine is when $A + 3B = 9$. The crucial geometrical fact here is that this represents a straight line. Points to the left and below the line (such as $A = 3$, $B = 1$) satisfy the constraint; those above and right fail it (such as $A = 1$, $B = 3$). It is the same with the stuffing-machine constraint, where the relevant straight line is $2A + B = 8$.

Drawing in the lines corresponding to these constraints carves out a shape called the feasible set: it is the region where the all constraints are satisfied. Every point within this region corresponds to a pair of values of A and B that the factory could produce within a day. This does not answer the question of what is optimal – after all, the point $A = 0$, $B = 0$ lies within the set, which corresponds to the factory being dormant. But it does address the equally important question of what is actually *possible*.

It is notable that all the lines here have come out straight rather than curved, which is the *linear* part of linear programming. This reflects the fact that all the quantities we have dealt with just involve As and Bs being added together, and thus we have avoided *non-linear* terms such as A^2 or $A \times B$. Out in the real world, non-linearity abounds of course, which makes it all the more remarkable – and convenient – that linear approximations are good enough for a large majority of optimization problems. (When they are not, the more complex field of non-linear programming can provide other techniques.)

It is the geometry of feasible sets that provides the central tools of optimization. In our toy factory example, the region has a finite area – good news (although it may not always be the case), since it guarantees that the problem will yield a solution. Another fundamental fact of linear programming is that the feasible set will always be convex, meaning it will have no holes or protrusions. To put it more formally, if one selects any two points within the set and joins them with a straight line, the entire line segment will remain inside the set – it will not need to pass outside. This is another piece of welcome news, since convex sets are geometrically far easier to handle than their non-convex cousins.

Geometrically, this is all sounding pleasingly tractable – but how much nearer are we to reaching our goals for toy production? The key to the whole thing is to understand the corners where the shape's edges meet, known as the extreme points of the set. The first really useful fact about linear programming is that the optimal value – if it exists at all – will always appear at one of these extreme points. This makes life immeasurably easier, in that the feasible region contains infinitely many points, but has only finitely many corners. For our bears and aardvarks there are just four, at (0,0), (0,3), (4,0), (3,2). So, to solve the problem, all we need to do is check which of these produces the maximal value of $A + B$. The corners would generate values of 0, 3, 4 and 5, so the solution to the problem is at the point $A = 3$, $B = 2$. It is with this allocation of resources that the factory will maximize its revenue, manufacturing a total of 500 toys per day.

Minimizing your pain

Of course, this is a rather simple example, chosen to convey the central ideas of linear programming. Because there were just two varying quantities (A and B), the toy factory generated a geometrical problem on the two-dimensional plane. In more realistic situations, as variables proliferate they generate vastly more complex feasible sets, which live in higher-dimensional geometrical spaces. Such shapes have the disadvantage of being impossible to visualize directly. Nevertheless, with the principles established, we still know that they remain convex shapes with finitely many corners. It is this satisfying truth that makes the esoteric-sounding study of 'multi-dimensional convex polytopes' a surprisingly practical topic.

Suppose a hospital is setting a roster for its nurses. There are various obvious requirements here. From a patient's perspective, the major criterion is that there should be sufficient nurses on duty to satisfy his likely needs, at every time of the day and night, each day of the week. The nurses themselves will have other concerns: neither too much nor too little work each week, adequate time off between shifts, and (typically) not too many night shifts. Where nurses have different specialities or grades, these will impose extra constraints. The target function to be maximized will be the overall extent to which nurses' working preferences are satisfied. It turns out that nurse rostering – or any equivalent task – is a deeply difficult theoretical conundrum (see *Teacher troubles*).

As a very different example, suppose the government of a small island wants to attract more holidaymakers. There are various projects they might decide to invest in, and these will attract tourists of different types: restoring historical temples will appeal to culture-hunters, while building luxury hotels and conference-centres will attract corporate interest, and new marinas will encourage the yachting fraternity and scuba divers. How to judge the right balance? Within each category, there might be an array of competing projects of differing costs and timescales. The target function here will be to optimize the estimated return on the investment.

Though superficially very different, when the constraints are worked out explicitly both of these scenarios will give rise to multi-dimensional convex polytopes – a mathematical testament to the many-sidedness of real life. The challenge comes not with the essential nature of the problem – which is easy to resolve in principle, as we saw with the toy factory – but with the scale of the numbers that the complexities of real life can throw up. If, say, we wanted to assign a schedule of 60 jobs to 60 people, who each have their own particular expertise and time constraints, the number of possible ways of doing this would $60 \times 59 \times 58 \times \ldots \times 2 \times 1$, more than the number of atoms in the observable universe! In realistic scenarios therefore, there may be many thousands of dimensions and constraints, carving out a feasible set with trillions of corners. To compute and compare the target function at all of them seems a wholly unrealistic demand.

Luckily, in 1947 the US mathematician George Dantzig developed a way to speed up the process beyond all reasonable expectation – the simplex algorithm. This procedure always seeks to optimize some target function sought by finding its *minimal* value. (There are standard tricks to convert maximization tasks into ones of minimization.) The process begins at a randomly chosen corner of the convex shape. The critical observation is this: if the target function is not minimized at that corner, then there must be an edge coming away from this point along which the target function decreases. So the simplex algorithm travels along that edge to the next corner, and then repeats the process. We can imagine the simplex algorithm as a small creature roaming the edges of the multi-dimensional polytope, visiting one extreme point after another, with the target function going down each time. When it arrives at a point where it can decrease no further, the minimum has been reached.

Dantzig's algorithm was a supremely elegant piece of mathematics. For moderately large problems, these techniques present a colossal improvement over a brute-force comparison

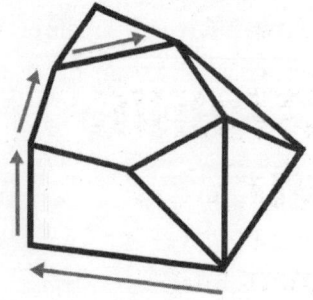

The simplex algorithm roams the edges of a multidimensional polytope, searching for the problem's optimal solution.

of all the extreme points. *Directly* comparing the relative benefits of all possible permutations is a practical impossibility, even for someone armed with the latest supercomputer. In contrast, decades of experience with the simplex algorithm strongly suggest that it can usually home in on the optimal solution in a number of steps comparable to the number of dimensions of the problem, keeping its running time within manageable proportions, even for large industrial problems. Nevertheless, the algorithm has been refined and augmented in the intervening years. The main variation has been the precise 'pivot rule' by which the next edge to be explored is selected.

Such a degree of time-saving efficiency has made the simplex algorithm one of the most popular and important algorithms in the world. Often it doesn't shout out its presence, but instead ticks away in the background, implemented billions of times a day by people and organizations across a broad range of industries and activities. All things considered, it is no exaggeration to say that George Dantzig's simplex algorithm is one of 20th-century mathematics' most important gifts to society.

OUR ELECTRONIC FRIENDS

The mathematics of
social networks

Humans have always been social animals. Not so long ago, our social networks were people we shared time with, face to face, or perhaps communicated with individually by letter or phone. That could be a large swell of people, or maybe just one man and his dog. That picture is already beginning to assume the sepia-tinged aura of a bygone age. With astonishing rapidity the very phrase 'social network' has been claimed by its online manifestations.

Today, it is estimated that over 10 per cent of the world's population – and rising – are participants in Facebook, MySpace, Twitter, Google+, Bebo and other social networking websites. As well as revolutionizing the way people keep in touch, online social networking is encouraging the development of a new type of mathematical sociology, spurred on by the fact that social networks are not simply multilateral communications channels; they also represent vast communities of potential customers, clients and target groups with shared interests. But how to estimate the size of that 'audience', and more subtly its degree of interconnectedness? And when, as happened in 2012, a leading social networking site is floating its shares, how might we evaluate its reach and influence?

Network laws

When we switch on our favourite television programme or radio station, our enjoyment is normally unaffected by the number of other people also tuning in. Of course, it may be that overall audience numbers *will* have an effect on what we see or hear in the long or medium run. We might enjoy a comedy show, but if few other people are watching, the station might not recommission the series – or might even drop it half-way through its run. But, in terms of our own response as a spectator of the show, the size of the audience is essentially irrelevant to us.

Sarnoff's Law, named after the early television pioneer David Sarnoff, extracts a consequence of this simple observation: that the total value of this type of network is *linearly* proportional to its size, so that if the number of users (viewers or listeners) approximately doubles or triples, so too does its value to its owners.

It is not the exact formula for the value that matters here, which will come down to the specific context. Mathematically, if n represents the audience size, it might be that the network's value is given by $22n + 71$ or $3n - 7$, or many of the other possible variations on this theme. What is important is the principle, this linear relationship, which is represented by the fact that while n appears in the formula, and there may be addition and subtraction involved, there is no more-complicated term such as a square (n^2) or a reciprocal $\left(\frac{1}{n}\right)$. Mathematicians often express this linearity in notation as the value varying '$O(n)$', spoken as 'with order n'.

An online social network is, of course, very different from a conventional television or radio network with its passive users. While we may not care if we are a television programme's only viewer, who would want to join a social network with no other users? The social network becomes increasingly valuable to us the more we can connect with our friends, the more we can pursue contacts for work, the greater the number of public figures we can follow, and so on. This is the definition of a network effect, a sense in which the whole is greater than the sum of the parts. But how *much* greater?

Various people have tried to quantify this effect more precisely, in terms of the value of the network. The most famous attempt came from the US electronic engineer Robert Metcalfe, the inventor of Ethernet network technology in the 1970s. He originally intended Metcalfe's Law to apply to small-scale physical networks, rather than the globe-spanning online giants we know today. The essence of his argument is that, as more machines are plugged into a network, its value increases as $O(n^2)$, so that if the number of users doubles, its value quadruples; if the number of users triples, its value increases nine-fold, and so on. Meanwhile, the cost of the outgoings needed to maintain this increased usership increases only as $O(n)$. In other words, once some critical threshold is passed, the benefits of expanding a network further are guaranteed to exceed the cost of doing so.

To appreciate this, it is useful to think about the possible number of connections among users rather than the total number of users. In a network of two people (A and B), there is only one possible connection. With a three-person network, there are three possible such connections: A&B, A&C and B&C.

With four people there are 6, with ten there are 45, and with a hundred there are 4,950.

The mathematical rule here is not complicated: in a network of n people – and assuming people do not connect to themselves – each may be connected to any of $n - 1$ others. But we also have to avoid double-counting connections: for example, A&B and B&A are the same thing. So the number of possible connections in an n-person network is obtained by multiplying n by $n - 1$ and dividing by 2, producing the equation

$$\tfrac{1}{2} \times n \times (n - 1)$$

A little algebra allows this formula to be rewritten as

$$\tfrac{1}{2}n^2 - \tfrac{1}{2}n$$

Again, it is not the algebraic details that matter here. The striking thing is the appearance of the term n^2, the *square* of the number of users. This is the justification for Metcalfe's Law, which, along with Moore's Law – the rule of technological progress, which says that the power of a standard computer doubles every two years – became a guiding principle of the 'dot com' boom of the 1990s.

Numerous refinements and alternatives to Metcalfe's Law have been suggested, with several writers suggesting that it is overly optimistic. The US computer scientist David Reed, however, turned the dial up even higher, arguing that in certain situations the value of a network can grow exponentially, which can be expressed mathematically as $O(2^n)$. Reed's Law is derived from counting neither the number of users, nor even the number of possible connections, but the number of sub-networks a single overarching network can support. Typically, these might correspond to different friendship or interest groups, or commercial, voluntary or charitable ventures. With a sub-network, each individual user has two options: to join or not to join. This means that if there are n users altogether, the total number of possible distinct sub-networks is 2^n (that is $2 \times 2 \times 2 \ldots \times 2$ where there are n 2s).

We can arrive at the same number another way: the total number of sub-networks is going to be equal to the total number of possible one-person groups (i.e. individual users) plus the number of two-person groups, three-person groups, four-person groups, and so on. It is a fundamental mathematical fact that adding up all these numbers will give 2^n.

Now, exponentials like 2^n are notorious for growing extremely quickly. If Reed's Law is correct, then the value of a network will *square* each time its user-base doubles. So if a network has an initial value of 100, then doubling in its user-base will increase the value to $100 \times 100 = 10,000$. A further doubling increases it to $10,000 \times 10,000 = 100,000,000$. Although Reed's Law is surely an exaggeration in terms of the financial value of the network, it does hint at the remarkable power and opportunities that modern online networks offer.

Small worlds and eccentric nodes

The interconnectedness of various networks has given rise to some entertaining explorations. The Kevin Bacon Game is a well-known pastime among movie buffs, who aim to connect actors to Kevin Bacon through a sequence of co-stars. Hugh

Laurie can be done in two steps; for example, he starred in *Plenty* (1985) with Meryl Streep, who in turn was Bacon's co-star in *The River Wild* (1994). Mathematicians have their own version of this game, begun in 1969 by Casper Goffman, and centred on the highly prolific researcher Paul Erdős (1913–96). By definition, he has an Erdős number of 0, while his co-authors (all 511 of them) have Erdős numbers of 1, and their co-authors (who number 6,593) have Erdős numbers of 2, and so on. Some mathematicians, such as those who have only written solo papers, have *infinite* Erdős numbers. But the largest known *finite* Erdős number is 13 – the furthest removed documentable link to Erdős – which is a rank held by five people. We also say Erdős has an 'eccentricity' of 13 within the collaboration graph. Among mathematicians who are mutually connected, the median (midway) and mode (most common) Erdős number is 5, while the mean average is 4.65. (For more on averages, see *The law of averages*.)

The two games have even been combined into Erdős–Bacon numbers, available only to select individuals who have authored scientific papers and appeared in films. The record of 3 is held by the mathematician Daniel Kleitman: he collaborated with Erdős and was a mathematical consultant to the film *Good Will Hunting* (1997), in which he also appeared as an extra, alongside Minnie Driver, Kevin Bacon's co-star in *Sleepers* (1996)!

The Erdős and Bacon numbers also illustrate the power of network-analysis to understand *how* people are connected. Because scientific papers and film casts are carefully indexed and archived, these sorts of collaboration networks are excellent test-beds for small-world studies, of which these games are just the most well-known examples.

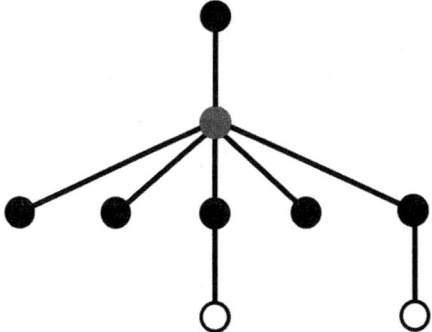

The grey node has an eccentricity of 2, the black nodes have eccentricity 3, and the white nodes eccentricity 4, giving this network a radius of 2 and a diameter of 4.

If we visualize the mathematical collaboration network laid out on a page – with a node for each researcher, and edges connecting pairs of co-authors – a node's eccentricity is the maximum distance from that node to any other. This gives a measure of how central the chosen node is within the network: central nodes will have high eccentricity, while those on the fringes will have lower values. Paul Erdős has an eccentricity of 13, as mentioned; but across the mathematical collaboration network as a whole the maximum eccentricity is 23, and is known as the network's diameter. The network's radius is the smallest value of the eccentricity for any node, which in this case is 12: somewhat surprisingly, there are two people, Izrail Gelfand and Yakov Sinai, who are more central to the network than Erdős himself!

In the geometry of circles, the diameter is precisely twice the radius. This is not the case with networks, but we can say something similar: that the diameter is *never more than twice* the radius. To navigate between two distant points, there is always the option of travelling from one to a central node, and then onward to the second, and each of these two trips cannot be shorter than the radius, at minimum.

Six degrees of separation?

What these games hint at is the suprising extent of interconnectedness in a highly populated planet. Frigyes Karinthy (1887–1938), a Hungarian writer and poet, is credited with first coining the phrase 'six degrees of separation' in 1929 to describe the average distance separating any two random people; in his day he meant handshakes of introduction (or suitable cultural equivalents). In full generality, the six-degrees conjecture is hard to verify, but in the 1960s Stanley Milgram's experiments with the US postal system suggested a US-only value of around six.

Mathematicians have been probing their collaboration network for decades, but the emergence of social networking sites has transported the study of degrees of separation into a much larger arena. In 2011, Reza Bakhshandeh and colleagues analysed the Twitter network and found that the average degree of separation between two randomly selected users was 3.4. A year before, a similar study found a value of 4.1. This decrease may seem surprising, since the network had certainly grown in the interim. But it reflects the network maturing: as it grows older, not only do new members join, but more importantly the existing members become more densely interconnected.

A milestone was reached in 2011 when a four-man team – Johan Ugander, Brian Karrer, Lars Backstrom and Cameron Marlow – analysed the entire Facebook network, amounting to around 721 million active users connected by 68.7 billion friendship edges. This put the mean-average friendship count at around 190 members, while the median was significantly lower at 99, the disparity caused by a small proportion of people with a very large number of friends (the maximum permitted being 5,000). Significantly, and reinforcing Karinthy's conjecture of almost a century before, was their discovery that 99.6 per cent of Facebook users worldwide were within six degrees of separation of one another, while 92 per cent were even closer, within five. These figures were bolstered by the finding that the mean distance between pairs of users is 4.7 worldwide, and closer, at 4.3, when limited to US users.

The extent of interconnectedness can be measured in subtler ways too. For instance, what proportion of pairs of a given Facebook user's friends are themselves friends? Ugander *et al.* found that for the median Facebook user, the answer was around 14 per cent, the number decreasing the greater the number of friends the central user has.

Intriguingly, the diameter and radius of the Facebook graph suggested by the data are actually infinite. The reason for this is that while 99.91 per cent of users form part of the main body of the graph, there are numerous detached components, of which the largest has just over 2,000 members.

Do size and interconnectedness matter – and the mathematical means for measuring those attributes? Certainly they did for Facebook owners and investors, when in 2012 the company was valued at a colossal $104 billion for its share flotation. In Karinthy's day, the six degrees of separation existed as a sociological or anthropological observation, but in a rather diffuse, spaced-out way – it described a network, by definition, but one that did not exist in any way that could be *used*. Today's online networks – mass subscription models, simultaneously accessed by interconnected millions – have created huge potential audiences, consumer bases and target groups.

Facebook's own line to potential advertisers, as per its website, is that 'People treat Facebook as an authentic part of their lives, so you can be sure you are connecting with real people with real interest in your products.' It is not all a one-way street of course; in the words of Jeff Bezos, founder of Amazon.com, 'If you make customers unhappy in the physical world, they might each tell 6 friends. If you make customers unhappy on the Internet, they can each tell 6,000 friends.' The exact relationship of social networking and commercial activity is a story still unfolding. But the mathematics of network size, connectedness and value will be central to it.

TEA-TASTING
AND TRIALS

The mathematics of
significance

Part of the appeal of mathematics is the satisfaction of producing a logically watertight proof, an argument that establishes a theorem with absolute certainty. Researchers in other areas of science generally do not have the luxury of tying things up so neatly, with no loop-holes or loose ends. The physical world is an all too messy place, plagued by coincidences, red herrings and blind alleys (see also *Smoke and mirrors*).

These traps may easily give the illusion of some new law or principle at work, and so scientists need to be self-scrutinizing. Before publicizing their results, they need to check the integrity of those results and rule out, as far as they can, the possibility that an apparently earth-shattering discovery or new wonder-drug is just a one-off event, a fluke. The primary tools for guarding against this eventuality come from the statistician's tool-box, in particular techniques for quantifying 'significance'.

Now: time to pause for a spot of tea.

Testing times

Around 1910, the biologist Muriel Bristol played an unexpected role in an important chapter in the history of science – not through her laboratory work on algae, but at an afternoon party with friends. This being England, the discussion soon turned to the topic of tea: in particular, was it better to pour the milk into the cup before the tea, or vice versa? Dr Bristol was adamant that she preferred tea with the milk added first, but another guest, Ronald Fisher, scoffed at this suggestion, arguing that there was no possible way to tell the difference. Being a gathering of scientists, it was clear that some sort of an experiment was needed to settle the matter. So, various cups of tea were prepared, in some of which the tea had been added first, and in others the milk. The results were unequivocal: Dr Bristol could indeed discern the difference.

Although this is certainly an impressive achievement, the real scientific breakthrough came later. The afternoon's events set Fisher's mind whirring. What would it take to settle the matter *rigorously*? After all, maybe Dr Bristol had simply guessed each time and got lucky. In other scenarios, such as medical research, the repercussions of a misdiagnosis would be far more serious than in the milk/tea conundrum. So Fisher set about designing what would become known as randomized tests to settle such matters. At their core are statistical considerations of a rather delicate nature.

To illustrate the principles, we may imagine that some medical researchers have come up with a proposed new treatment, Drug X, for a minor medical problem, Condition Y. To test it, they find some volunteers suffering with Condition Y, apply Drug X, and then watch to see whether or not they improve. So far, so simple, but there is plenty that can go wrong. It may be that the results show some sufferers improving, while others remain ill. But how can we decipher this mixed picture? Might the subjects who improve have been on the mend anyway? How can the researchers know?

In this scenario, the researchers need to decide between two competing hypotheses. The 'null hypothesis' is the technical term for the default position: in this case, it will state that Drug X has no impact on the health of patients with Condition Y. The 'alternative hypothesis' is that Drug X *does* have an effect. For Dr Bristol's teatime experiment, the null hypothesis would be that she was unable to discern the order of ingredients with any more accuracy that blind guessing, the alternative hypothesis being that she had some ability to do so. In each case, the primary rule of significance testing is that the researchers *must accept* the null hypothesis *unless* there is enough evidence to reject it in favour of the alternative.

Returning to our lab, let's suppose that the researchers test 20 sufferers with Drug X, and find that 9 of them recover within a week. This looks promising for Drug X, but from previous investigations the researchers might also know that 25 per cent of people with Condition Y recover within a week untreated, on average. Occasionally, then, among a group of 20 sufferers it will happen that 9 or more recover by chance, in which case Drug X may actually have played no role. The mathematical question to be answered is: what is the chance of getting a result as extreme as $\frac{9}{20}$ purely by chance if the drug has no effect – that is, if the null hypothesis is true? This critical number is known as the experiment's *p*-value. Calculating the *p*-value is central to the process of hypothesis testing, which proceeds broadly as follows:

- decide on a *level of significance* depending on the size of the experiment and the level of rigour needed – perhaps 5 per cent, 1 per cent or even 0.1 per cent;

- calculate the experiment's p-value;

- reject the null hypothesis if – and only if – the *p*-value falls below the significance threshold (which would enable us to say that the drug works or that Dr Bristol does indeed have a remarkably refined palette).

If the *p*-value is higher than the significance threshold, the null hypothesis has to be accepted. This does not *necessarily* mean that the drug doesn't work or the tea-taster has no talent; it might simply mean that the significance level selected was too high a bar for the test carried out, meaning that more tests are needed.

The idea is not too complex, but the technical mathematics arises in step 2, when it comes to working out the experiment's *p*-value. In our drug-testing example, it comes out at around 4.1 per cent (see box overleaf), which means we could reject the null hypothesis at the 5 per cent level of significance. Can we declare that the drug works? That would depend on the context. While that level of testing would be all well and good for an afternoon tea party, medical testing normally demands a higher level of statistical rigour, so researchers might well want a result at the level of 1 per cent or 0.1 per cent significance, in which case they would embark on further tests; but they could certainly do so with some optimism.

Thrashing out thresholds

There are various ways of calculating the crucial p-value, depending on context. For our drug trial and tea experiment, the appropriate method involves a mathematical device known as the binomial distribution. In Dr Bristol's case, assuming she had no ability to discern milk first/last, her chances of guessing correctly would be 50 per cent. With a little historical revisionism, we can imagine her presented with eight cups of tea, in each of which the order of ingredients is decided by the toss of a coin – and she calls them all correctly. The p-value answers the question: what is the chance of her achieving this through blind guessing? She has odds of $\frac{1}{2}$ of guessing one cup correctly, and for guessing two correctly the odds are $\frac{1}{2} \times \frac{1}{2} = \frac{1}{4}$. For guessing three cups correctly, the odds are:

$$\frac{1}{2} \times \frac{1}{2} \times \frac{1}{2} = \frac{1}{8}$$

And so on, so her chance of guessing all eight cups correctly can be expressed as

$$\left(\frac{1}{2}\right)^8 = \frac{1}{256}$$

This produces a figure of around 0.004. This p-value is comfortably within the 1 per cent (and indeed the 0.01 per cent) significance level.

What, however, if Dr Bristol guesses seven out of eight correctly? This is trickier to handle. So let's address a more specific question first: what is the chance that she will guess the first two cups correctly, the third incorrectly, and the remainder correctly? At each of the eight steps there is a 50 per cent chance of the appropriate answer, so the answer is again $\left(\frac{1}{2}\right)^8 = \frac{1}{256}$.

Now, there are eight possibilities along these lines, depending on whether she calls the first, second, third, etc., cup wrongly. Each possibility has the same probability of $\frac{1}{256}$, giving a total of:

$$8 \times \frac{1}{256} = \frac{1}{32}$$

In other words, around 0.03.

In statistical tests, the p-value represents the chance of getting a result *at least as strong* as the one we do obtain. Here, it means the chances of Dr Bristol guessing seven or eight cups correctly. Therefore, to the value of $\frac{1}{32}$ we add on the previous value of $\frac{1}{256}$, giving a p-value of around $\frac{9}{256}$, or around 0.035. Since this is still below the 5 per cent (0.05) significance limit, Dr Bristol would pass the test at the 5 per cent level of confidence.

Blind trials and blind alleys

In our talk of drug testing we have skipped over one mysterious but hugely important factor: the placebo effect. It is now well-known that treatment that has no medical efficacy can nevertheless produce improvements in a wide-range of medical conditions. This means, for a testing regimen, that it is not enough simply to show that Drug X is more effective than doing nothing; it must also prove itself better than a placebo.

To grapple with this, medical testing typically involves an experimental group and a control group. Both will be given tablets or injections (or whatever form the drug takes), but only for the experimental group will the medication be Drug X. For the control group, it will be a sugar pill or saline injection. The statistical test will then compare the two groups' performance.

The placebo effect is, of course, a strange and poorly

DRUG X: A CASESTUDY IN BINOMIAL DISTRIBUTION

For our notional drug trial, we need the all-important *p*-value, the chance that at least 9 of the 20 sufferers of Condition Y would recover anyway within a week. Our given information is that the chance of one chosen patient recovering naturally within one week is 25 per cent (0.25), and the binomial distribution can use this to tell us the *p*-value.

We begin by specifying nine people in advance. The probability that they will recover is 0.25^9, while the possibility that the remaining 11 subjects will not recover is 0.75^{11}, making the overall likelihood that those nine – and only those nine – recover $0.25^9 \times 0.75^{11}$. The next ingredient we need to know is the number of possible combinations of nine people that we might pick from 20.

The answer here may be succinctly expressed using factorials (see *Hold the line, please!*), which involve multiplying together all the whole numbers up to some limit. For example, $5! = 5 \times 4 \times 3 \times 2 \times 1$. It turns out, in our drug-trial example, that the number of possible ways of picking 9 people from a group of 20 is:

$$\frac{20!}{9! \times 11!}$$

(See *Hold the line, please!* for an explanation of this figure.) Putting this together, the chance of exactly nine people recovering is

$$\frac{20!}{9! \times 11!} \times 0.25^9 \times 0.75^{11}$$

This produces a figure of around 2.7 per cent (0.027). For the purposes of the statistical test, however, it is more appropriate to calculate the chance of *at least* nine people recovering by chance. This means adding up the equivalent expressions for 9, 10, 11, 12, …, 20.

Doing this produces the overall *p*-value of around 0.041, or 4.1 per cent.

understood phenomenon, and so for a *fair* experiment it is crucial that test subjects do not know which group they are in: it is a blind test. To reinforce this discipline, it is now standard practice for even the medics administering the treatment not to know whether they are providing a particular subject with Drug X or the placebo. This double-blind testing is the cornerstone of modern medical research.

Whether in drug testing or any other investigation, however stringent the statistical test may be, there is always some probability that conclusions will go awry. The experiment may produce a 'false positive', causing the researchers to reject the null hypothesis when it should be accepted – judging that Drug X has an impact when it does not, or incorrectly crediting Dr Bristol with unwarranted tea-tasting skills. The likelihood of this is precisely given by the significance threshold, so if that were set at 5 per cent, then the test would admit a 5 per cent chance of a false positive.

By contrast, to accept the null hypothesis when it should be rejected is a 'false negative', and this type of error is much harder to legislate against. If Drug X does have a definite impact on Condition Y, beyond the placebo effect, it may be so small as to be missed in a clinical trial, generating a false negative. Or our tea-taster may get it right 60 per cent of the time, definitely better than blind guess but only slightly, and again this may not be picked up in a small-scale test. This problem of false negatives relates to one of the most misunderstood topics in the whole subject: the meaning of *significance*.

When is significance significant?

To describe a finding as 'statistically significant' seems to suggest, well, that the result is significant in the sense of *mattering* – but it may be nothing of the sort. It is a distinction that regularly causes confusion among media commentators, and even among scientists themselves.

We can see this from the probability theorist's favourite example: tossing a coin. Any real coin, if examined closely enough, will likely have some design feature or construction defect that *very slightly* skews its probabilities. It might be that the likelihood of getting heads is 50.01 per cent, and that of tails is 49.99 per cent. In nearly all respects, this is not a 'significant' difference, and it will be missed by simple experiments. If we toss the coin 10,000 times and get 5,001 heads, this gives a p-value of 49.6 per cent, nowhere near the 5 per cent threshold (the detailed argument is similar to the binomial test of Drug X). Technically, this is a false negative, for the test leads us to accept the coin as fair when it is not.

For all practical purposes, this matters little; the coin is so close to being fair as makes no difference. Yet, with enough time and effort, it would be possible to run an experiment establishing the coin as unfair. Suppose a machine tosses the coin 100 million times and gets 50,010,000 heads. This gives a p-value of 0.023 (again using a binomial test). So, we now have firm evidence of bias, at the 5 per cent level of significance. With further effort this could be improved to 1 per cent, 0.1 per cent – or in fact any desired level – although the tests will take an increasingly long time.

The distinction is between the actual result being *scientifically* significant, and the evidence for that result being *statistically* significant. These two are completely separate conclusions, a fact that is not widely appreciated and which generates misunderstandings. It might be that having five minutes of extra sleep per night decreases our likelihood of developing certain diseases, and it may be this could be established to a level of statistical significance. (We can imagine the headlines: 'Live longer with just 5 minutes extra sleep!') But the real scientific question is not about the statistical significance of the test; it is whether the health benefits are substantial enough to worth bothering with.

The economist Milton Friedman said that 'the only relevant test of the validity of a hypothesis is comparison of prediction with experience'. Certainly, in the end, the mathematics does not stand in for measured judgements, born of experience, about what really matters. Nevertheless, whether in the lab or at the tea-party, an essential pre-requisite for making such judgements is to get to grips with the statistics of significance.

THE CALM AT THE EYE
OF THE STORM

Fixed points and the nature
of equilibrium

It is a tenet of Buddhism that everything changes, whether we like it or not. To Buddhists and non-Buddhists alike, that sense can be all too palpable. The yearning for peace and equilibrium, in the midst of situations that feel fragile and undependable, is a feeling that many of us have experienced. Perhaps this is why, in the face of life's vicissitudes, human societies have placed a high value on the ability to keep a steady hand on the tiller. It is a sentiment captured most famously in the opening lines of Rudyard Kipling's paean to stoicism, *If*: 'If you can keep your head, while all about you / Are losing theirs and blaming it on you ...'

Islands of stability, amid broader scenes of confusion, have also provided enduring subjects of fascination for mathematicians and physicists, in a variety of contexts. Indeed, around the beginning of the 20th century a whole new field was born, topology, whose practitioners searched for constants among the properties of shapes, features which endured no matter how much those shapes might be mangled and deformed. Surprisingly – and encouragingly for those who bemoan a world in flux – various thinkers began to discover fixed-point theorems and related findings that identified well-behaved, stable locations, even in the midst of the most confusing situations. These lines of enquiry have gone on to permeate a wide variety of discourses, from fundamental physics to questions about how human societies and economies function.

Coffee and doughnuts

The Hungarian mathematician Alfréd Rényi famously summed up a mathematician as 'a machine for turning coffee into theorems'. But he did not have in mind the direct inspiration that the Dutch philosopher–mathematician L.E.J. Brouwer (1881–1966) found in his own cup. As he sat pondering profound mathematical issues, Brouwer gently stirred his coffee with a spoon, and then let it settle again. It hardly seems one of history's greatest scientific experiments, yet Brouwer extracted a surprising conclusion from this simple situation: that there must be at least one point within the drink that returns to exactly the same location that it occupied before the stirring.

He went on to elaborate his celebrated Fixed-Point Theorem, which in turn spawned a slew of related results, with implications throughout mathematics and beyond. However, the information Brouwer's result provided was limited: it did not tell him *where* in his coffee cup the fixed point could be found – but it did assert that it must exist somewhere.

Brouwer contributed other examples of fixed points in the physical world. If we take two identical pages from the same issue of a newspaper, lay one out flat on the floor and scrunch the other crudely into a ball and toss it on top of the first sheet, Brouwer's theorem guarantees that – so long as the ball lies within the borders of the flat page – there will be some point on the scrunched-up page that sits exactly over the corresponding point on the flat page. So Brouwer's theorem applies even when the geometrical manipulation is complicated; while the liquid in the coffee cup was just rearranged, our newspaper is deformed in a more complicated way, with the result that there may be several points of the scrunched-up page over a single spot on the flat page.

The simplest instance where Brouwer's theorem applies is a segment of straight line. If the points along this line are rearranged, Brouwer's theorem guarantees that there will be some fixed point at the same position as in the original line. This example is fairly easily graspable, for example by taking two tape-measures, laying one flat (the line segment) and folding the other as many times as desired and placing it carefully on top, so it does not overlap the first. There is guaranteed to be some number on the folded tape that is directly above its twin on the other tape. Even if the folded tape is also stretched or compressed, the same fact holds true. The fine print is that that rearrangement must be *continuous* – it is easy to violate the rules by cutting the tape in half and swapping over the two sides; this will not leave a

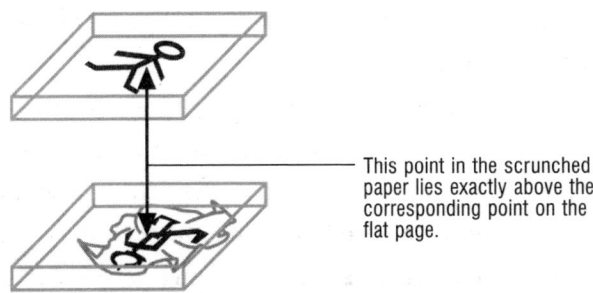

This point in the scrunched paper lies exactly above the corresponding point on the flat page.

fixed point. More subtly, it is crucial that the endpoints of the line must be included in the experiment; to see why, imagine water flowing through an infinitely long pipe. If all the liquid flows 6 inches forwards, then again we will find no fixed points.

A natural question must be: do all shapes generate fixed points in this way? This can be answered by imagining Brower's coffee cup remoulded into a doughnut shape – technically known as a torus (see *The hole story*). To contain the coffee one has to imagine a circular trough with a hole in the middle. In a mug of this design, a light stir would circulate the coffee around the central hole, and there is no reason why any of the liquid should settle in the same place: it might all just circulate clockwise or anticlockwise by a few degrees. There may be no fixed points at all.

Brouwer's theorem is therefore not a universal observation; it applies to some shapes but not others. It does apply to any one-dimensional line-segment, two-dimensional disc or three (or higher) dimensional ball, so long as the edges are included in the shape and the rearrangement is continuous. What is more, being a *topological* result, it applies to any shape that can be pulled into any of these forms, such as a rectangular page of newspaper which is equivalent to a disc, or Brouwer's cup of coffee, which is topologically a three-dimensional ball.

This essentially means that any geometrical figure that contains no holes (unlike the torus) and whose boundary is included (unlike the long pipe) satisfies the conditions.

'You are here'

A special case of the kind of deformation we have discussed applies in the case of maps. Mathematicians use the words 'map' and 'mapping' in a variety of technical senses, but for most people a map (of, say, their home city) is simply a shrunken version of the larger reality – as it were, the city in miniature. In 1922, Stefan Banach made an important discovery in regard to miniaturization. If a shape is shrunk in such a way that the miniaturized version remains inside the original – which is to say the map of the city is physically placed within the city limits – then there will again be some point that remains fixed and stable during the shrinking process.

Banach's shrinking result dispenses with some of the requirements of Brouwer's original Fixed-Point Theorem. To start with, it does not matter whether the shape contains any holes (although its edge still needs to be included). What is more the conclusion is tighter: Brouwer's theorem guaranteed the existence of at least one fixed point. Banach's tells us that there will be *exactly* one. Most importantly, though, this special location is now easy to find. On a map, the fixed point marks the map's current location within the city – the 'you are here' spot. If you place a pin there, then the location of the pin within the real city is exactly described by its own position on the map. No other city location satisfies this condition.

The shrinking result is known as Banach's Contraction Mapping Theorem. It does not require that the shrinking need be uniform – though that would normally be the case in a map done to scale. Deformation is permitted, even to extreme levels. The only requirement is that every pair of points within the shape has to end up closer together than they were in the original city.

Does the fixed point remain if the shrinking process occurs not once but many times? It can be tested out with a familiar party trick, by pointing a video camera at a television screen displaying the camera's image. The effect is an image repeated *ad infinitum*, apparently receding down a long tunnel. Gradually the sequence of retractions shrinks down to a single point, the light 'at the end of the tunnel'. That is the unique fixed point, and its location exactly coincides with its image on every one of the nested television screens. It can be confirmed by marking a dot at that point on the screen. While there will be numerous outlines of televisions, there will be only one dot.

Earthly stillness

It may, so far, appear that this interest in fixed points is just a rarefied parlour game, what with its doughnut-shaped coffee cups and tricks with newspapers and cameras. All very interesting, but does it matter? Well, one consequence is that we can rest assured that, at this exact moment, there is some point on the surface of the earth where there is absolutely no wind at all. That conclusion is the unavoidable consequence of another of Brouwer's seminal discoveries, the famous Hairy Ball Theorem. It asserts that that if you take a spherical object covered with hair – a coconut may well come to mind – and try to comb it all flat, you are guaranteed to leave a crown or tuft somewhere. If we imagine each strand of hair as representing the wind direction at that point on the globe, then a crown or tuft will be the point of stillness – perhaps even the eye at the centre of a storm.

This theorem was an early result in topology, and the topological criteria mean that it does not just apply to coconuts or the earth but to any subject that can be pulled or twisted into the form of a sphere. It need not apply – again – to shapes that do not

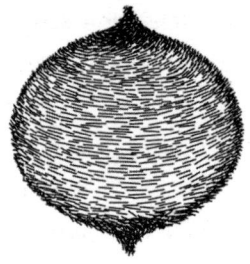 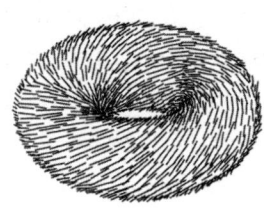

Unlike a sphere, hair on torus can be combed flat, without leaving any tufts or crowns.

fit that definition, and once again the doughnut-shaped torus is an exception. A hairy bagel may be a slightly disgusting thought, but it would be possible to comb its hairs perfectly smooth. The rings of Saturn, for example, may be permanently windy at every location.

The mathematician Heinz Hopf investigated the 'comb-ability' or otherwise of a broad variety of shapes, and his Index Theorem of 1926 lifted Brouwer's analysis into mathematical hyperspace– where shapes with more than the three dimensions come into play. So, while the familiar two-dimensional spherical surface of a coconut cannot be combed flat, its higher-dimensional brother, the 3-hypersphere *can* be combed. A simple hairy circle – also known as a 1-sphere – can be combed too, but the 4-hypersphere cannot. In general, the answer is as beautiful as it is unexpected: hyperspheres of odd dimension can be combed without leaving crowns, while those of even dimensions cannot.

Competition and equilibrium

Fixed-point theorems can provide insights into unexpect- edly diverse phenomena, and have found a particular role in describing points of equilibrium in competitive and market scenarios. Indeed, such are their attraction that two mathematicians, Piet Hein and John Nash, invented a board game called Hex that exhibited the power of fixed-point theorems in action.

The appeal of Hex is its simplicity. It is designed for two players – one with black counters and one with white – and its name derives from the design of the board, which is divided into hexagonal cells. The board begins empty, and the players take turns to place their counters in cells. From the outset each player owns two opposite edges of the board: say black owns north and south, while white owns east and west. Each player's aim is simply to build a bridge linking their two sides, from counters of the correct colour.

One advantage of Hex over some other games (from Noughts & Crosses to Chess) is that it is guaranteed to produce a win for one or other player: a player can only be sure that he cannot win if his path from one side of the board to the other has become completely blocked, and that only happens when the other player has completed a chain and won. To phrase this fact as a mathematical theorem, we might say that every possible assignment of black or white counters on the board will include a winning chain for either black or white (but not both). Mathematical statements such as this one require a rigorous mathematical proof, no matter how obvious they may seem, and in this case Brouwer's Fixed-Point Theorem is exactly the necessary ingredient. In fact, the result runs the other way too: from the assumption that any game of Hex will terminate, it is possible to deduce Brouwer's Fixed-Point Theorem for a two-dimensional disc.

Higher-dimensional versions of the game can also be

concocted (though they are not so easy to set up). The three-dimensional version is playable on a computer, and involves three people taking turns placing counters in three-dimensional cells, each aiming to build a bridge from north to south, east to west, or top to bottom of the playing space. This game is also guaranteed not to end in a draw, a fact that is equivalent to Brouwer's theorem for a three-dimensional ball.

Brouwer's theorem arises in game theory in another way, through the classic pastime of Scissors–Paper–Stone. Here, two players count to three and then choose, via conventional hand gestures, one of the three objects. If the players pick the same item, the game is a draw. Otherwise scissors beats (cuts) paper, which beats (wraps) stone, which in turn beats (blunts) scissors. Usually the game is a way to break a deadlock in light-hearted decision-making; but if we take the game more seriously, one might want to seek out the best tactics for playing it. The answer will depend on how the opponent is playing. If we notice that she is playing Scissors more often than the other two, we should increase the frequency with which we pick Stone, and decrease Paper. But she might respond by increasing the frequency with which she plays Paper. A subtle strategic battle may ensue . . . or not.

Suppose our opponent picks totally randomly each time. Now there is *no* strategy we can adopt that will exploit any weakness in her position, and our best plan would be to choose randomly as well, so to, at least, not give her anything to latch onto, and in turn no reason to change her strategy. Then we and our opponent would be locked into what game theorists call an equilibrium.

In 1951, John Nash proved that every game within a broad class (not limited to two-player games) is guaranteed to have an equilibrium somewhere – a situation where no player has any incentive to change strategy, even with complete information about the other players' tactics. Nash first derived this important fact in his PhD thesis, as a consequence of Brouwer's Fixed-Point Theorem. It became a centrepiece in the growing subject of game theory, and contributed to his winning the Nobel Memorial Prize in Economics, in 1994.

Nash's result, and the many variations on the theme that have subsequently been discovered, continue to fascinate. They are valuable for analysing the frameworks in which people make decisions, where the possibility of gain must be weighed against the dangers of loss, and where one's own best strategy will depend on the behaviour of other people (see *The rise of* homo economicus). So the concept of a Nash equilibrium can illuminate the more familiar idea of a 'stalemate', and provide real insights into political or diplomatic deadlocks. The concept is also now standard terminology within economics, where a market is defined to be in equilibrium at that idyllic moment when prices for all the goods are set so that supply and demand match exactly.

As L.E.J. Brouwer sat stirring his coffee, surely he could never have expected the wide-ranging consequences that his insights would bring.

ONE SMALL STEP . . .

The mathematics of
space travel

By any measure, one of humanity's most spectacular achievements came in July 1969, when Neil Armstrong and Buzz Aldrin became the first people to set foot on the moon. It was a mere twelve years after the Soviet Union had sparked the space race with the launching of Sputnik, the first earth satellite, and it provided the millions of people around the world who watched the images on their television screens with a sense of living through a moment of enormous historical and scientific significance.

A hidden debt for that slice of history is owed to Richard Arenstorf, who, six years previously, performed a piece of mathematical analysis that ultimately made the moon landings possible. Put simply, he provided the route map for the journey that Apollo 11 took. But, to do this, he first had to grapple with one of the most notorious questions in the history of mathematics – the so-called three-body problem.

The laws of planetary motion were first set down in the 17th century by Johannes Kepler, and subsequently explained by Isaac Newton (see *The dynamic solar system*). Of central importance was Kepler's discovery that planets orbit the sun not in circles but in ellipses. Newton accounted for this fact by analysing the gravitational interplay between the planet in question and the sun. But what would happen if a third object were introduced into this scenario – say the sun, a planet, and a moon orbiting that planet? This three-body problem turned out to be incomparably more difficult than Kepler's two-body version.

A problem exceeding the human mind

When considering two bodies in space – call them A and B – there are only two forces involved, the gravitational effect of A on B and B on A. Introducing a third body increases the number of forces to six: each of A, B, C on the other. With two bodies, there are essentially three possible outcomes, depending on the strength of the force. The two can be drawn together until they collide, they might cycle around each other in elliptical orbits, or the two might hurtle away from each other. (Technically this will happen along a path described by a parabola or hyperbola, curves closely related to ellipses but with infinite length, which never loop back to connect with themselves as ellipses do.)

With the introduction of a third body, the situation becomes much more complex. To start with, in the two-body problem the geometry is essentially two-dimensional. Although the planet and the sun travel through three-dimensional space, their initial directions of movement define a flat two-dimensional plane. As the bodies cycle around each other, the system is guaranteed to remain within this special plane. But this is not true for the three-body problem, whose objects will typically travel through all possible dimensions.

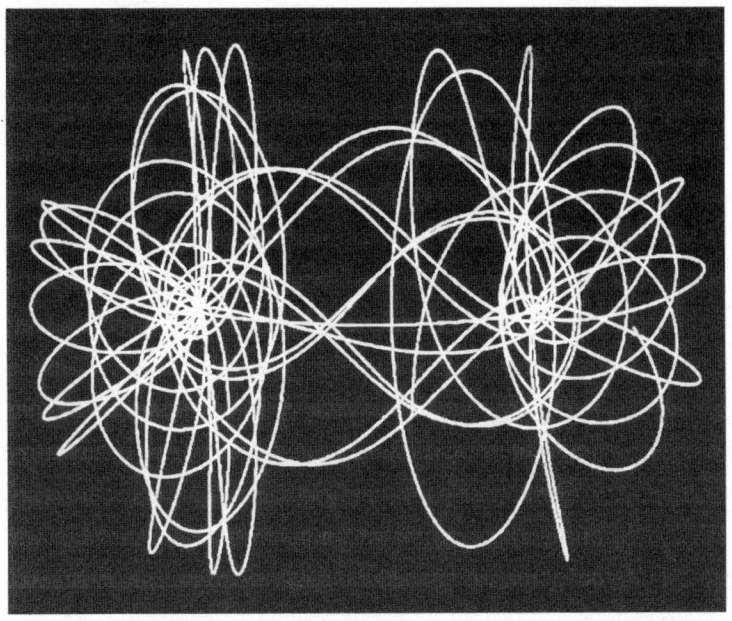

J.C. Sprott's model of the trajectory of a planet orbiting two static stars of equal mass, illustrating the chaos inherent in the three-body problem.

But the difficulties run deeper still. The two-body problem is characterized by periodicity, that is, a planet orbiting a star will trace out the same path, over and over again. Typical three-body problems do not yield this regularity; in fact, they are liable never to repeat themselves at all, but trace out fiendishly complicated curves through space, defying any simple description. This troublesome topic was studied by several of the greatest minds in mathematics, including the Swiss-German Leonhard Euler in the 18th century as well as his predecessor, Newton, who had given up on the task, declaring that 'to define these motions by exact laws admitting of easy calculation exceeds, if I am not mistaken, the force of any human mind'.

In 1887, there was renewed interest in the three-body problem, from an unexpected quarter. King Oscar II of Sweden, himself a mathematics graduate from the University of Uppsala, decided to offer a prize of 2,500 Swedish Crowns for a solution to the greatest open question in the subject. The mathematician Gösta Mittag-Leffler advised him that the three-body problem was the one to pick, and a new generation of geometers set their sights on the problem, among them the Frenchman Henri Poincaré. Although a complete analysis continued to elude those who took up the challenge, Poincaré was able to make considerable progress, such that he was able to claim King Oscar's prize.

More than that, Poincaré's work on the three-body problem proved to be the launching pad for a whole new intellectual field – chaos theory. He realized that even the tiniest modification to the initial state of the system – say increasing the weight of the moon by a grain of sand, or shifting the planet away from the Sun by a millimetre – could produce an entirely different outcome in the long run. (More familiarly we know this principle as the 'butterfly effect', the term coined in the early 1960s by Edward Lorenz: see *Rain or shine?*)

Just when the mathematical community had despaired of completely cracking the three-body problem, the astronomer Karl Sundman astonished the world with a solution, which he published in a series of papers between 1906 and 1912. Sundman was a reserved man, who had studied astronomy at observatories and universities around Europe before taking a lecturing job at the University of Helsinki in his native Finland. Starting with data describing the initial positions and speeds of the three bodies, Sundman was able to provide a formula, of sorts, which would accurately describe the state of the system at every subsequent moment.

Sundman's work was a true *tour de force*. Although he missed out on King Oscar's prize, now that it had gone to Poincaré, he was awarded the Pontécoulant Prize by the French Academy of Science, which even doubled the usual prize money in recognition of the extraordinary nature of his achievement. Notwithstanding all this, however, Sundman's work by no means closed the book on the three-body problem. The reason was the mysterious and highly unusual nature of his solution.

The ordinary elliptical orbit of a planet in the two-body problem is described by a short equation. Sundman's solution to the three-body problem was altogether more fiendish, taking the form of an infinite series. This means that the bodies' locations are described by what appears an impossible procedure – mathematical expressions in which infinitely many quantities need to be added together. An example of how a series might begin would be:

$$a_0 + a_1 s + a_2 s^2 + a_3 s^3 + a_4 s^4 + \cdots$$

In this expression, the numbers represented by a_0, a_1, a_2, etc. are fixed quantities calculated from the initial data about the system: the starting speeds and positions of the three bodies. The number s represents the cube root of the time t that has passed since the clock was started.

But how does an infinite series help – since adding together infinitely many numbers is self-evidently impossible? In fact, scientists are often faced with this sort of conundrum, and the usual work-around is to add together the first few terms, maybe ten, perhaps a hundred, possibly even (with the power of modern computation) a million, depending on the required level of accuracy. Almost always this will provide a good enough approximation. Unfortunately though, this will not work with Sundman's formula. His series converges exceptionally slowly, and extracting any useful information out of it would mean adding together an unfeasible number of terms, in the order of

$$10^{10^8}$$

that is, a 1 followed by 100 million zeroes!

A cushion against chaos

At this point, we should perhaps remind ourselves of the observable facts of our solar system. There is a very obvious objection to the wild numbers and unpredictabilities outlined so far: the actual three-body problem that *we* inhabit – the sun, the earth and our moon – bears little resemblance to the chaotic and near indescribable situation Sundman analysed. On the contrary – and for the human race, this is a blessing – our own three-body problem is tame and predictable.

The explanation for this is that there is a region of stability around any planet, known as its Hill Sphere, named after the US astronomer–mathematician George William Hill (1838–1914). Inside our Hill Sphere, the movement of any moon or satellite will be dominated by the gravitational attraction of the earth, and we can safely ignore the effect of the sun. Technically, the radius of the Hill Sphere is given by the expression $a \times \sqrt[3]{\dfrac{m}{3M}}$.

Here, a is the radius of the planet's orbit around the star (assuming, for simplicity, a circular rather than elliptical orbit). In the case of the earth, this is around 150 million kilometres, or

15×10^7 km. Meanwhile m represents the mass of the planet – in the earth's case, around 6×10^{24} kilograms, and M is that of the sun, approximately 2×10^{30} kilograms.

Plugging these numbers into the formula produces a Hill Sphere around the earth of a radius measuring around 1.5 million kilometres. Since our moon is only 0.4 million kilometres away from us, it is comfortably within this region. The upshot is that it is safe to model the sun–earth–moon relationship as two 2-body systems: sun–earth and earth–moon.

Lest we get too sanguine about our stable environment, we should remember that any trip to the moon necessarily brings three objects into play. We can ignore the effects of the sun, thanks to Hill, but we have to contend with a different kind of third object: the spacecraft. However, in this case there is a way to simplify the mathematics, by observing that one of the bodies that is involved – the spacecraft – is considerably lighter than the other two. Ignoring the tiny gravitational effects of the spacecraft on the earth and moon reduces the number of forces involved from six possible ones to four: the earth and moon pulling on each other, and each pulling upon the spacecraft. This simplification has a further consequence. As with Kepler's two-body problem, all the motion can again be assumed to take place within a flat two-dimensional plane, defined by the motion of the earth and moon.

Although simpler than the full-blown three-body problem, this restricted version is still very far from straightforward to resolve, despite several attempts being made over the centuries. A complete mathematical solution to this problem would entail a geometrical description of every possible motion of a spacecraft.

Mathematically, this is an unfeasible task. Luckily, from the point of view of space travel a complete solution is not required.

Conveyer belt to the moon

This brings us back to Richard Arenstorf and the mission to put a man on the moon. The whole project was undertaken in the tense political environment of the Cold War space race, with much national pride staked by the United States and the Soviet Union on their respective achievements. Politicians on both sides did not miss the fact that the space race allowed a spectacular but peaceful demonstration of a fearsome technological prowess, which, if the need arose, could be turned to more deadly ends. The Soviet Union could claim the early successes – not only rocketing Sputnik 1 into orbit in 1957, but also putting the first unmanned craft on the moon when Luna 2 crashed into the lunar surface in 1959. US politicians were determined that they should claim the ultimate triumph of a manned trip to the moon, a promise made public by President Kennedy, committing the United States to invest time and resources in solving the many difficulties involved – not least, those of the underlying mathematics.

It was in this context that Richard Arenstorf, a mathematician working at NASA, became the latest in a long line of thinkers to turn his attention to the three-body problem. Arenstorf focused on the restricted, two-dimensional planar version of the problem, and what he sought was not a complete solution, but rather a *stable orbit*. Just as planets follow elliptical orbits around a star, so a satellite on an Arenstorf orbit would travel around both the earth and the moon. In 1963, he was able to

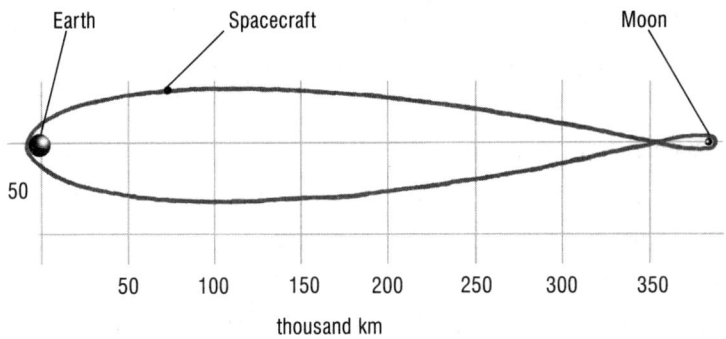

Apollo 11's route to the Moon, as provided by Richard Arenstorf's orbit.

find a family of such paths. It turns out that an Arenstorf orbit resembles a figure of eight, with the moon and earth inside each of the two loops. These orbits are stable and periodic – that is, they repeat themselves – and they avoid the chaos associated with the general three-body problem. Arenstorf even envisioned a 'space bus' to travel continuously on the orbit, much as an ordinary satellite circulates around the earth.

The space bus is for the future – but Arenstorf orbits have comprehensively been demonstrated to be the right routes for lunar expeditions to take. A craft taking off from earth can use its rockets to hop onto the orbit. If it does so correctly, pointing in the right direction and travelling at the appropriate speed, gravity will then do the rest of the work. It only remains for the craft to hop off again at the moon.

What is more, by performing some delicate calculations on an early computer, Arenstorf was able to tweak the numbers to bring the orbit very close to the surfaces of the earth and moon, and thus to make take-off and landing as efficient as possible. In 1966, Arenstorf was awarded the Medal for Exceptional Scientific

Achievement by NASA, in recognition of his accomplishments. It is no exaggeration to say the discovery of the Arenstorf orbital route was the mathematical breakthrough that made Apollo 11's historic trip to the moon possible.

'One small step for man, one giant leap for mankind' summed up the astronauts' triumph in 1969. Mathematics too took a giant leap in that era, with Richard Arenstorf's victory over one of the most notorious problems in the subject's history. But that was not the last time that humans would have to grapple with the three-body problem, nor was it the last time that mathematical progress would take us further into space. Unmanned craft, including NASA's two Voyager craft have gone far further than the moon, flying past Jupiter, Saturn, Uranus and Neptune, and now exiting the solar system altogether – defying the gravitational attraction of the sun. They manage this by flying close to other planets, picking up speed, which then flings them away from the sun. This is another trick made possible by a careful mathematical analysis of the three-body problem.

TULIP BUBBLES AND HEDGE FUNDS

Futures, options and the ups and downs of markets

As the financial crisis and credit crunch of 2007/2008 took hold, it was not just the holders of US sub-prime mortgages whose fortunes tumbled. The reputation of the banking industry as a whole fell to its lowest point in modern history, its glittering financial models and exotic products looking as threadbare as the Emperor's new clothes. Among the artefacts of banking practice to attract the most opprobrium were financial derivatives, the complex instruments blamed in many quarters for fuelling an unsustainable credit bubble.

The financial models that traders use to price commodities and products (including debts) are undeniably complicated. Nevertheless, the essential ideas behind derivative trading are surprisingly simple. They are also of long standing, dating back almost 400 years to one of the most unusual events in economic history: the Dutch tulip bubble.

Flowers of fortune, hedges against risk

Several decades before the 1630s, the tulip had begun to captivate the affections of Europeans, after samples were sent back from the

Ottoman Empire. These flowers became luxury items and status symbols – and their value and price increased. New varieties were developed and imported, and specialist tulip traders began to make serious amounts of money.

Such fashions come and go of course, but there was no precedent for the way in which the Dutch tulip market soared between November 1636 and February 1637. Prices rocketed to ever more outrageous heights, until single bulbs were trading for more money than a craftsman might hope to make in ten years. But not everyone who was buying tulip bulbs wanted flowers to decorate their homes; people scented profit, and were prepared to sell land and other valued possessions to buy bulbs, in the expectation of selling them on for even greater sums. In February 1637, it was all over: the tulip bubble burst, with the value of bulbs plummeting by over 95 per cent, destroying the fortunes of many of those who had invested at the wrong time.

It was not only the extraordinary extent of the Dutch tulip bubble which distinguished this episode from other rises and falls of the market, but also the technical details of the financial instruments used: early forms of derivatives that continue to be traded around the world today, the commonest of which are futures and options. They derive from the desire to insure oneself against the unknowable events that lie ahead and, for those who think they can predict what the future holds, to attempt to profit from that knowledge.

To illustrate the principles, a sideways step from tulips to rice is appropriate, rice being another of the oldest established derivates markets in the world. If we imagine that restaurant-owner Alison wants to buy rice from farmer Brian, the two might come to an agreement that, in exactly three months time, Brian will sell Alison 100 kilograms of rice at a price of £1 per kilo. The contract they have agreed is a 'future'. Why would the two enter into this? It might be that Alison doesn't need the rice yet, and she will be better able to manage her budget for other goods if she knows how much rice she will get, when, and at what cost. Similarly, it might be that Brian has no rice to sell until his harvest comes in; he too might benefit simply from the certainty of the deal to be done. There may be other practical reasons. There is mutual advantage in that the agreement allows both farmer and restaurateur not to worry about the price of rice in the intervening months; it may go up or down without affecting them. In this scenario, Brian and Alison are said to have 'hedged' their exposure to the market.

Hedging risk is not the only motivation for trading futures. It might be that one party, or both, sees an opportunity to profit. This will depend on what each thinks will happen to the price of rice in the time that passes between their striking the deal and the goods finally changing hands. If Alison thinks that the price is likely to rise above £1 per kilogram in the next three months, then she is certainly better off purchasing a future from Brian than buying rice at market price on the day. Indeed, she might not even want any rice; her hope just might be to immediately sell it on for £2 per kilogram, thus making £100 profit. In this scenario, she is said to be speculating on the price of rice.

Brian may also be speculating, but if so he will be hoping that the price of rice falls. If the market price in three months time is 50 pence per kilogram, then he is better-off under his agreement with Alison than he is selling it on the day. Again, he may not even have any rice of his own to sell; his plan might be to buy it on the day for 50 pence per kilogram, and then sell it to Alison at their agreed price, immediately finding himself £50 in profit.

An 'option' differs from a future, in our example, when Alison is no longer committed to buying the agreed quantity of rice at the fixed price (known as the strike price) on the agreed date. Rather, she secures the right to buy it if she wishes, while Brian is forbidden from pulling out of the deal. What can Brian possibly gain from this scenario? Seemingly nothing; so he will only agree to the deal if Alison pays him a fee – a premium – for doing so. For her, the worst case will be to lose the premium, and then abandon the rest of the deal.

In fact, more precisely this deal is a 'call option'. The alternative is a 'put option', whereby Brian would secure the right to sell to Alison a fixed amount of rice, at a fixed price, on a fixed date – but not the obligation to do so. In this version, the boot is on the other foot, and Brian would have to pay Alison a premium, while she is forbidden from pulling out of the deal.

Tulip mania and tulip misery

Back in the heady days of 1636, trading in tulip futures had become widespread. This was a natural development for a seasonal product, since bulbs could only be dug up and exchanged during the summer. The rest of the year, with the bulbs in the ground, transactions involving actual tulips could not take place. In the autumn of 1636, an astonishing idea was floated, which was ratified by the Dutch courts the following year. It stated – to use modern terminology – that all tulip futures contracts would be automatically converted into call options.

Rather than being committed to buying the bulbs at the agreed price, buyers could pull out of the deals and instead pay a small forfeit or fee in compensation to the seller, which was set at 3.5 per cent of the price.

This (with hindsight) disastrous move fuelled a final mad flurry of tulip trading. Buyers were happy to agree to ever more ludicrous prices, lured by the potential for profit, while reassured by the provision of this new safety net. Meanwhile, sellers suddenly felt at risk; if they were only likely to receive 3.5 per cent of the asking price, then that price had better be very high.

Then catastrophe struck – the bottom fell out of the tulip market. Speculators stopped seeing the likelihood of making any money, and preferred to cut their losses and pay the forfeit. As deals fell through *en masse*, no-one was prepared to pay the previous high prices. Sellers who had taken huge financial risks to buy into the market were left holding plant bulbs of little value.

Optional mathematics

Mathematics has been central to trade since the days of bartering, but modern mathematical finance had to wait for over 250 years after the tulip crash for its awakening, at the dawn of the 20th century. When it finally came into bloom, options-pricing was one of the chief objects of study. Louis Bachelier applied his mind to considering the terms and conditions that buyer and seller might agree on. Returning to our example, if Alison wished to buy a call option from Brian, then what premium should Brian charge her? And what should Alison be prepared to pay? If, a month down the line, Alison wanted to sell her option to a third party, what should she charge them?

Before Bachelier could tackle these questions, he needed some way to describe the unpredictable ebb and flow of the underlying stock – the price of rice, or tulips, or whatever it might be. He developed a novel approach for doing this, which, when rediscovered in a very different context, would subsequently become known as a 'Brownian motion'.

'Brownian motion' was named in honour of Robert Brown (1773–1858), who wondered what forces governed the seemingly random movements of tiny fragments of pollen grains, suspended in water. This question would later be answered by Albert Einstein who produced a mathematical model of a particle in water, being buffeted to and fro by billions of water molecules every second. Subsequent experiments by Jean-Baptiste Perrin would show that Einstein's mathematical model was an excellent fit to reality; this triumph was the final decisive piece of evidence for the fact that matter is composed of molecules. It is a source of justified pride to financial mathematicians that Bachelier independently arrived at Einstein's model before the physicists!

Bachelier's model begins with a term for Brownian motion: B_t, where t represents time. As t changes, B_t moves up and down at random. Now, B_t alone does not represent the price of the stock. In a Brownian motion, in the physical sense, the particle is taken to be at the point 0 to start with before its movement is analysed. But in the context of market trading, the stock will not begin with price of 0. To correct for this, we add on the stock's initial price, that is, its price at time $t = 0$. We can refer to this value as S_0.

Some stocks vacillate wildly, while others are more constant; this is quantified by a number called the *volatility* (v). Multiplying B_t by v will either magnify or dampen the extent of the price changes. If we put all this together, the price of the stock (S_t) at time t is modelled as the initial price (S_0), give or take some random change (B_t) and magnified by the volatility (v), and produces the equation:

$$S_t = S_0 + vB_t$$

Now, on the basis of this model, what can we say an option is worth? The answer is the difference between the market price (S) on the day of reckoning (which we might call time T) – so that is S_T – and the agreed strike price (K). How much should someone pay for an option? The answer is: anything up to $S_T - K$. The problem is, of course, that this is unknowable before time T is reached. Luckily, Bachelier's formula above allows the price S_T to be estimated in advance.

The critical notion is that of expected value, which roughly means the average of all possible prices, weighted by likelihood (see *The rise of* homo economicus). So, Bachelier argued, the correct price for an option is the expected value of $S_T - K$. What is more, a deep analysis of Brownian motion allowed him to provide an explicit formula for this number. With this done, traders could enter the relevant parameters (volatility, starting price and time remaining on the option), solve the resulting equation, and decide whether or not an option was likely to represent a good deal. Of course, this is necessarily an *average* result: sometimes it will be too low, and sometimes too high. On average, though, it stacked up well.

Bachelier's model was later superseded in financial circles by one developed in the 1970s by US academics Fischer Black and Myron Scholes, which became the industry standard. Their approach was essentially the same: to begin with a Brownian motion and use it, along with a measure of volatility, to model the price of the stock; then calculate the expected value, and finally derive a formula for the price of an option. The only difference is technical. Where Bachelier's model was built upon a standard Brownian motion (B_t), Black and Scholes used a geometric Brownian motion, of the form e^{B_t}, invoking the revered constant e (approximately 2.71828), a number that figures prominently in mathematical accounts of growth.

It is a subtle difference between the two models: an ordinary Brownian motion changes in an additive way; in other words, to calculate the change over two months, we add together the changes in each month. In Black–Scholes's geometric model, we multiply them, much as interest rates are multiplied (see *The law of averages*). The alteration allowed extra economic background to be built into the model, such as the underlying interest rate.

The dangers of derivatives

The Black–Scholes model remains the standard way to trade options today. For the most part, it has accorded well with experience; at any rate, many people have made a great deal of money with it. At the same time, the 21st-century's financial crisis has placed the whole apparatus and mechanism of financial speculation under the spotlight as never before. A concern is that derivatives are meta-products; put simply, they are bets on the price of the underlying stock – or, even further removed, bets on those bets. But the expansion of derivatives markets means that they have taken on a life of their own and their value now dwarfs that of the stocks from which they are derived. In 2010, the quantitative analyst Paul Wilmott estimated the notional value of the international derivatives market at US $1.2 quadrillion, 200 times the world's annual gross domestic income!

The writer Nassim Nicholas Taleb is one of those who have been ferociously critical of the Black–Scholes formula and related tools. The assumption he attacks is the bedrock of financial mathematics: Bachelier's Brownian motion. While this may describe the day-to-day fluctuations of a market satisfactorily, it fails to account for the impact of large-scale events and unforeseen disasters, from the bursting of bubbles (tulips in 1637, the US housing market in 2007) to man-made and natural catastrophes such as the 9/11 attacks and the Japanese tsunami in 2011. Taleb dubs such one-off, game-changing events 'Black Swans'. When missed by the model, they may plunge the markets into a state of confusion.

Mathematical models, used wisely, may enrich an individual, a company, a nation. But old-fashioned caution about understanding the limits of the equations, and about the unpredictability of life, remains an essential – if sometimes forgotten – bedrock of finance. In the words of Emanuel Derman and Paul Wilmott's *Financial Modelers' Manifesto*, published online following the 2007–8 financial crash, 'Our experience in the financial arena has taught us to be very humble in applying mathematics to markets, and to be extremely wary of ambitious theories, which are in the end trying to model human behavior. We like simplicity, but we like to remember that it is our models that are simple, not the world.'

TEACHER TROUBLES

The tricky world
of timetables

Being in the right place at the right time is generally acknowledged to be a fortunate state of affairs. It may provide the lucky break that launches a glittering career. At a more mundane and literal level – though equally handy for one's career prospects – it is the daily routine for schoolchildren, university students and their teachers and professors, and it is encapsulated in the humble timetable. Although usually taken for granted, that simple grid of lessons and activities conceals multiple complexities – hurdles to be overcome by the administrator charged with bringing it into being, by balancing subjects, year groups, classes, teachers (full-time and part-time), rooms, equipment, budgets, extracurricular activities and much else – while attempting to keep a cool head. The task is fraught with difficulty, and the challenges run so deep as to generate some staggering numbers, and feed into the biggest single question in theoretical computer science.

In terms of mathematical genres, timetabling is an example of a scheduling task, of the type that typically arises in transport networks, in manufacturing, and across organizations and educational institutions. In building a car, for instance, some tasks can be done in parallel – such as assembling the engine and doors – but others must be done in a particular order. It's not a good idea to attach the wheels before the driveshaft is fitted, for instance. This observation helped inspire Henry Ford's creation of the moving production line. A school's timetable may, conceptually, appear one of the simpler scheduling tasks, since it only entails the avoidance of clashes rather than doing things in the correct order. But this emphatically does *not* translate into a quick way to complete it.

To a mathematician, timetable-construction can be translated into the seemingly very different terminology of networks. We begin with a list of tasks to be done – perhaps the lessons to be taught in a single day, say Year 7 Science, Year 8 Science, Year 7 English, etc. – and represent each task by a node on the page. The central issue is that certain tasks cannot be carried out simultaneously, an occurrence indicated by drawing an edge (or line) between such pairs. So the node representing Year 7's English lesson with Mr Hardy will be connected to the one representing the same year's Science class with Mrs Robinson, which in turn is connected to (since it clashes with) Year 10's Science class also taught by Mrs Robinson, and so on. This example, so far, consists of three nodes where A is connected to B which is connected to C, but C is *not* connected to A.

With such a diagram of nodes and edges drawn, the next thing to do is to add a splash of colour. In fact, we are going to paint each node, choosing the colours cleverly so that no edge connects two nodes of the same colour – the idea being that the colours will represent different time periods within the timetable. So, red nodes might stand for the hour 9–10am, while green represents

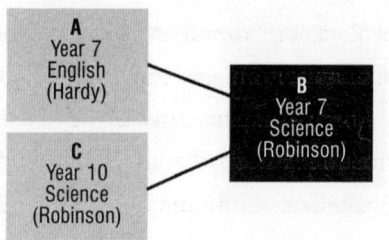

This small, three-node network has a chromatic number of 2, since A and C each clash with B, but not with each other. For larger networks, this value is hard to calculate.

10–11am, and so on. So long as we do not connect nodes of the same colour, we have avoided time clashes. But we have also arrived at a major question: what is the smallest number of different colours needed? For mathematicians, this minimum number of colours is the network's chromatic number, and for our school it represents a fundamental piece of information: the minimum number of timeslots needed to complete all the tasks. If the chromatic number exceeds the periods available in the school day, then this particular timetabling task is impossible! But, assuming that is not the case, once the chromatic number is determined, there may well be several ways to complete the colouring – there may be many, equally good, timetabling outcomes.

A universe-worth of calculations

So far, this talk of nodes and colours suggests a diverting pastime involving crayons, pencils and rulers and perhaps the promise of a sudden revelation as the relationships slot into place. Alas, it is not quite like that. Mathematicians have studied tasks like this intensively in recent decades, and the result is bad news for the school timetabler. Finding the chromatic number of a network is called, in the jargon, an 'NP-complete problem', which means that it is a difficult task, which takes a long time – and in general there is unlikely to be any quick shortcut.

Of course, crayons and rulers bit the dust some years ago, and one might ask, in the computer age, how difficult is 'difficult' and how long is 'long'? Surely the answers to these questions must depend on the chosen software, the speed and power of the computer, the ingenuity of the method used? All that is perfectly true, but there is a more profound notion of a task's difficulty, which is completely objective and even largely immune to technological progress. This falls in the category of computational complexity, a topic studied both in university mathematics departments and the research laboratories of software companies. The idea is to quantify the difficulty of a task according to the minimum number of steps needed to complete it – the fewer the steps, the easier the task is said to be.

The size of the network obviously matters, and the critical question here is how quickly the computation grows relative to that size. Research so far suggests that a general network with n nodes will require around 2^n steps (that is n-many 2s multiplied together) to locate its chromatic number. It is somewhat intimidating news, since this expression grows very fast indeed. When $n = 10$, it means our computation will require $2^{10} = 1,024$ steps; but doubling the network size to 20 nodes will mean a computation of $2^{20} = 1,048,576$ steps, while a 30-node network involves 2^{30} steps, which is more than a billion. It is easy to see these computations exploding out of control. At this rate, a network containing 90 nodes (or, equivalently, a timetable of 90 different lessons) would demand a computation taking longer than the lifetime of the universe to run, even computing one step every nanosecond. Such is the devastating power of exponential growth.

To P or not to P – the polynomial question

All may not be lost, however. Most practical computer programs come from a far friendlier class of algorithms with polynomial growth. To illustrate this we might imagine that a brand new way for analysing a network is discovered, which requires n^2 steps to calculate the chromatic number of a network with n nodes. This would actually constitute a gigantic speed-up. For a network with 10 nodes, our program will require $10 \times 10 = 100$ steps. By the time we have $n = 90$ inputs, we have only reached $90 \times 90 = 8,100$ steps, which most would agree is a considerable advance on the age of the universe: a modern computer processor could do it in the blink of an eye.

To the trained eye, the speed of these two algorithms can be seen in the contrasting algebraic expressions – the (exponential) 2^n and the (polynomial) n^2. Other examples of polynomials are expressions like n^3, n^4, n^5. These algebraic tags allow us to classify tasks according to difficulty. It is known that there are many tasks that *can* be solved in polynomial time. The collection of all such polynomially-solvable tasks is traditionally known simply as P. A helpful (if somewhat simplified) rule-of-thumb is that P consists of all those tasks for which it is possible to come up with an algorithm fast enough to be useful in the real world.

What would be a boon for our original timetabling challenge would be an algorithm that could complete the job quickly, meaning in polynomial time. Does it exist? The theoretical question here is whether or not the timetabling task lies within the polynomial category – is it 'in P'? The bad news is that the answer, so far as we know, is 'no'.

There is, though, a close cousin of the timetabling task

that *can* be done quickly – namely, checking an answer. If we contemplate a 90-node network corresponding to a timetabling problem, and we are presented with a colouring of it using just eight colours, it would be fairly quick work to check whether or not the offered solution delivers the goods. All our task consists of is looking at each edge in turn and checking that it does not connect two nodes of the same colour. Tasks like this, which can be *checked* in polynomial time, are said to be 'in *NP*' (technically standing for 'non-deterministic polynomial time').

And there's the rub. The single biggest conundrum in theoretical computer science is whether or not every task that is 'in *NP*' must also be in '*P* '. On the face of it, there is an obvious answer: 'no'. And it is true that a majority – though not all – of computer scientists do indeed believe that *P* cannot equate to *NP* in this way. Indeed, it is sometimes said that if mathematicians were to adopt the standards of proof accepted in other sciences, they would long since have accepted this as a law of the universe. After all, for over 50 years since the Austrian-born mathematician Kurt Gödel first hinted towards this question, many talented scholars have analysed it in great depth, chief among them Stephen Cook, Leonid Levin and Richard Karp, whose analysis in the early 1970s elevated this question to the status of one of the greatest in the whole of mathematics. The fruit of this research is that there seem to be certain tasks – timetabling among them – that are in *NP* but not in *P*.

But mathematicians operate to a different, more stringent level of rigour. It is not enough simply to observe that timetabling *seems* not to lie in *P*. What is needed is a rock-solid proof that it does not. As yet, that has not been forthcoming, either for timetabling or for any other *NP* task.

In fact timetabling is special. Not only is it in *NP*, but it is *NP*-complete. This means that it is maximally difficult among all *NP* tasks, and if any *NP* task lies outside *P* (i.e. is definitely not polynomially computable) then timetabling must do so too. The converse to this is that the humble school timetabling problem has an unexpected significance: if anyone can find quick way to crack it, in full generality, it will rank as one of the greatest and most unexpected achievements in modern mathematics, since it will automatically follow that he or she has stumbled upon a proof of $P = NP$.

Suck it and see – a mathematical workaround

One would be forgiven, at this point, for asking how it is possible that any school on the planet actually functions, if its timetable epitomizes a mathematical conundrum of such monstrous proportions. Certainly few of those charged with compiling a timetable would wish to pit themselves against the twin terrors of exponential growth and *NP*-completeness. The truth is that, while there is no mathematically perfect solution, there are computer-aided workarounds, which make individual incarnations of the problem tractable.

To start with, it may not actually be necessary to find a graph's chromatic number. There are, after all, a fixed number of periods in the school day, say seven. There is no advantage in squeezing the entire timetable into six sessions, even if that turned out to be possible, so we don't need the best possible answer, just one that is good enough.

Secondly, *NP*-completeness applies to the problem in full generality – a method for finding the chromatic number of *any network whatsoever*. In particular instances, there may be regularities in the network that make the problem easier to solve. For instance, it may be that every class has a broadly similar number of lessons, and that the same applies for teachers. This gives the network some overall symmetry, which might be exploitable – by a technique known as a genetic algorithm.

In recent years, genetic algorithms have become popular practical tools, which originated in research into artificial intelligence (see *Creating electronic brains*). They do not loop around searching for the optimal answer as a traditional computer program might. Instead, they make guesses and then try to refine or adapt them, thus gradually 'feeling' their way to a workable overall solution. Such a system might colour a few nodes of the network at random, and then try to expand these coloured regions, avoiding clashes, making changes, or backtracking when required. The term 'genetic algorithm' comes from an analogy with evolution, which takes place through dual processes of random mutation ('the guess') followed by natural selection ('the refinement'). Their approach is also strikingly similar to how a human might attack the problem, though the computer has the advantage of added speed – as well as never getting bored or frustrated. In a curious way, we are almost back with our crayons and rulers again, trying things out.

Today, genetic algorithms are used in all manner of cutting-edge applications, from code-breaking to artificial creativity – the production of original art or music by computer. Yet among these sexy manifestations of 21st-century technological prowess there is also the seemingly drab school-timetable problem, a deceptively mundane task, but one with truly remarkable hidden depths.

LET THERE BE LIGHT

The extraordinarily useful
geometry of optics

From the cosmically distant to the microscopically close, in the 21st century we can see more, and in greater detail, than ever before. It is no exaggeration to say that the optical lens and curved, magnifying mirror represent two of humanity's greatest technological breakthroughs. They have reconfigured our perception of ourselves, of the world around us and of the universe beyond. The science of optics has a myriad of applications, from the supremely practical to the very edges of science, and underpinning them all are a few fundamental geometrical ideas.

The lens itself was not a single discovery. 'Reading stones' were rudimentary lenses in the ninth century, and the general phenomenon of man-made magnifying lenses dates back to the Ancient Assyrians of around 1000 BC. But it was not until the early European Renaissance that optical technology began to make serious advances, with the first eyeglasses being created in Italy around the year 1300. The application of the lens to scientific observation had to wait until the early 17th century, when the new inventions of the microscope and the telescope played a pivotal role in the scientific revolution and in establishing the heliocentric worldview (see *The dynamic solar system*), while polymaths such as Isaac Newton investigated the properties of light. And with Newton's further experiments with unusually shaped mirrors, he sowed the seeds for the technology of modern astronomy.

To create such optical appliances – the ancestors of today's most advanced microscopes and telescopes – first meant understanding the geometry behind the reflection, refraction and the focusing of light.

Reflection and the power of curved mirrors

The principle behind a conventional flat mirror is quite simple: that light reflects off its surface at the same angle from which it approaches, but in the opposite direction. So a laser beam that strikes a mirror at 30° from the left will then bounce off at 30° to the right. In this respect there is nothing distinctive about light – other things being equal, it operates in the same way as soundwaves or, for example, a ball bouncing against a wall.

A device such as a flashlight, however, takes advantage of a more exotically shaped reflector behind its bulb: a paraboloid. On its own, the flashlight's bulb would radiate light in all directions, so it needs a mirrored background to reflect all the light forwards. Ideally, all the rays of light should come out parallel, forming a nicely directed beam rather than scattering. This is what a paraboloid will ensure.

Defined geometrically, a paraboloid is the collection of all points whose distance from a fixed spot is the same as that from a particular flat plane. In a flashlight, that fixed spot, the focus, is where the bulb is positioned. While the flashlight

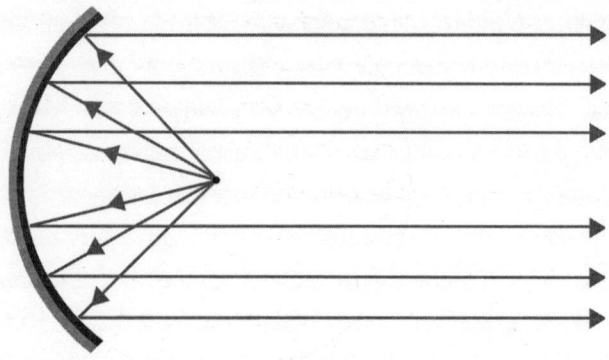

A bulb at the focus of a parabolic mirror will produce perfectly parallel rays of light, a very useful fact in optical engineering.

is a technological newcomer, the same parabolic shape has fascinated geometers since the Ancient Greeks. One particular observation has proved enduringly important: that a horizontal line will strike any point on the surface at the same angle as a straight line connecting that point to the focus. In terms of light, this guarantees the flashlight's backing will reflect all the light into parallel beams running from left to right.

The same principle applies, in reverse, in radio-telescopes and satellite dishes, which also adopt a parabolic backing surface. This time, radio waves are incoming parallel beams, and if the dish is properly aligned so they that hit it face on, they will all be reflected towards the focus. It is at this sweet spot that the sensor sits.

Refraction, distortion and all the colours of the rainbow

While reflection takes place when light bounces off surfaces, refraction occurs when it passes from one medium into another.

It is demonstrated by the familiar experience of gazing at objects through water, and seeing them appear somehow distorted, or seeming to sit in the wrong place. A bath plug will appear misplaced because the light has slightly deviated as it passes from the water into the air on the way to our eyes.

The reasons for – and degree of – distortion derive from the fact that light does not travel at the same speed through different media. The usually quoted value for the speed of light is 299,792,458 metres per second, but this describes light travelling through a vacuum. In water it is significantly slowed down – by around 1.33 times – and in optical-grade glass it travels slower still, by a factor of around 1.62.

These numbers, 1.33 and 1.62, are known as the 'refractive indices' of water and glass respectively. Every medium has its own refractive index, and mathematically it is defined as: $\frac{c}{v}$ where c is the speed of light in a vacuum, and v is its speed through the medium in question. So the refractive index is the proportion by which light is slowed down. Crucially, a material's refractive index also determines the degree to which the direction of a beam of light changes as it enters the medium.

An important advance was made in the 17th century by the Dutch astronomer Willebrord Snellius. Snell's Law, as it became known, established the fundamental rule of refraction. (In fact, it was a form of rediscovery, the same rule having been observed in the 10th century by the Persian scholar Ibn Sahl.) To grasp Snell's Law requires a couple of critical geometrical concepts.

The first is the notion of the *sine* of an angle. If we imagine a 1-metre-long pole held at an angle of $X°$ to the horizontal, then sin X is exactly the vertical height of the pole's end from the ground. Thus, sin 0° = 0 metres and (should the pole be held on the ground completely upright) sin 90° = 1 metre. In other words, we can consider the value of the sine function as measuring how close the angle is to being perpendicular. (For more on the sine function, see *Wave-worlds*.)

Snell's Law is expressed in terms of the 'normal' to a surface, understood as a line emanating from it at exactly 90°. In optics, when a beam of light strikes the surface of a medium such as water or glass, it is the angle between the beam and the normal that is critical. A beam hitting the surface front-on by definition makes an angle of 0° to the normal. (Indeed, such a beam would not deviate from its previous path as it entered the medium, but continue straight on.)

To understand Snell's Law, suppose that a laser beam hits a bathtub of water, making an angle of θ_1 to the normal (θ, the Greek letter *theta*, is the letter traditionally preferred by geometers to denote angles). On the other side of the surface – here that means underwater – the ray continues at an angle of θ_2 measured against the underwater normal. The relationship between θ_1 and θ_2 determines how much the ray deviates from a straight line at the point of entering the water. If $\theta_1 = \theta_2$ then the beam does not deflect at all (see diagram opposite).

This relationship between θ_1 and θ_2 is given by Snell's Law. It connects the angles θ_1 and θ_2 to the refractive indices of the two media, call them n_1 and n_2, and it asserts that

$$\frac{\sin \theta_1}{\sin \theta_2} = \frac{n_2}{n_1}$$

Thus, to plug in some figures, if our beam of light has hit the water in the bathtub at 45°, and the refractive index of air is

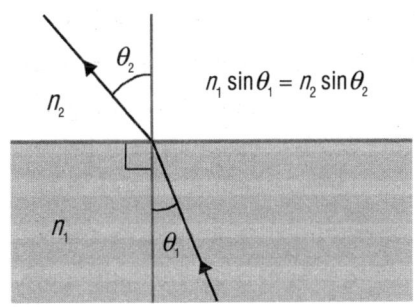

Snell's law describes refraction, by relating the angles θ_1 to the refractive indices of the two materials, θ_2.

$$n_1 \sin\theta_1 = n_2 \sin\theta_2$$

approximately 1 while that of water (as mentioned above) is 1.33, then Snell's Law tells us that

$$\frac{\sin 45^0}{\sin \theta_2} = \frac{1.33}{1}$$

Doing all the calculations, the answer for θ_2 comes out as approximately 32°. Meanwhile, the *difference* between θ_1 and θ_2 will indicate how much the beam deviates from its path, which in this case is around 13°. And this last figure quantifies the illusion of the misplaced bath plug.

Snell's Law, as we shall see, has proved hugely important. Optical technology relies on understanding how light refracts through different media – especially glass. Yet it is not the whole story of refraction, for thus far we have skipped over something: the angle by which light refracts depends not only on the media through which it is travelling, but also on the wavelength of the light. In our calculations above, the quoted figures are for yellow light with a wavelength of around 589 nanometres. (There are 1 billion nanometres in a metre.) It is a fact of nature that light with a shorter wavelength, such as violet (around 400 nano-metres) refracts less than light with a longer wavelength, such as red (at approximately 650 nanometres).

It is this fact that Isaac Newton exploited in his famous experiment of refracting white light twice through a glass prism in 1672, so that the different light components, with their different degrees of refraction, dispersed into a rainbow. In fact, Newton was not the first to realize that a glass prism – or for that matter a rainstorm – could have this effect on white light. But, before his investigations, it was believed that in some mysterious way the glass (or water) was somehow colouring the light. Newton's first corrective was to observe that when pure red light passed through a prism, it emerged unchanged: so much for the prism's supposed colouring ability. But the real triumph came when he was able to use a lens and a second prism to refract the rainbow back into a single beam of white light. At this stage the conclusion was unavoidable: white light is not some colourless, 'pure' version of light; rather, it is the result of mixing together light of every colour of the rainbow.

Lens lessons

It was Newton's century that witnessed a revolution in the understanding, design and application of the optical lens. By the later Middle Ages, reading glasses existed as everyday aids, but in 1590 the Dutch spectacle-maker Zaccharias Janssen had the idea of packing several lenses together into a tube. His idea would prove transformational, as telescopes enabled distant bodies to be studied in unimagined depth, and for the first time microscopes brought scientists face to face with the world of the very small.

The effect of a lens depends on a twofold refraction – once when light enters the lens and once when it leaves. By carefully

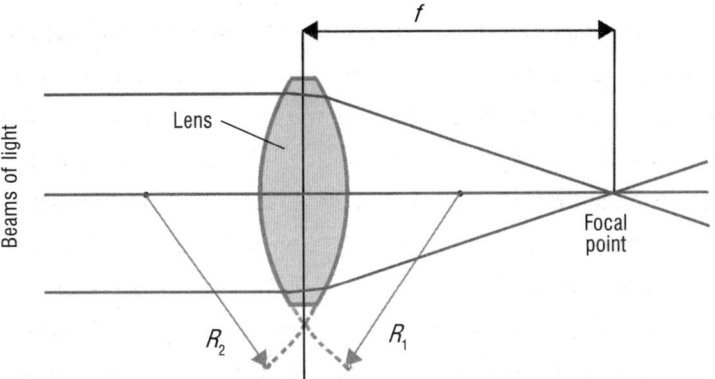

The lens-makers formula relates the geometry of a lens to its optical strength.

curving the two sides of the lens, this double refraction can focus parallel beams of light at a particular point. The *strength* of the lens is defined by the distance from the lens to this focal point (f). The nearer the focal point, the stronger the lens, and the smaller the value of f. The lens-maker's formula, discovered by the 17th-century philosopher René Descartes, relates this crucial distance to three other numbers: the refractive index of the lens material, which we might call n, and two measurements from the geometry of the lens itself.

Geometrically, the formula considers lenses whose curved faces form parts of spheres, and for simplicity we will assume we are dealing here with a convex lens, in which both sides bulge out, though the same idea works for concave lenses. The remaining two measurements are the radii of these spheres, R_1 and R_2. The lens-maker's formula then states that:

$$\frac{1}{f} \approx (n-1) \times \left(\frac{1}{R_1} + \frac{1}{R_2} \right)$$

(The symbol $\approx$ means 'approximately equals'.) This formula enables detailed calculations on how to design a lens correctly to produce a particular effect. It also lets us read off some familiar properties of lenses. The more steeply curved the two sides of the lens, the smaller the values of R_1 and R_2 will be, and so the smaller f will be too, giving a correspondingly stronger lens.

Reflecting on space

The refractional properties of lenses informed the first telescopes that emerged in the early 17th century, including that developed by Galileo, which combined a convex and a concave lens. But Newton's experiments also produced the first functional *reflecting* telescope, employing curved mirrors rather than lenses. His design was similar to the parabolic satellite dishes of today, but placed within a tube. At the focus, instead of a modern electric sensor, was a second flat mirror angling the image upwards to the eyeball of an observer peering down into the tube.

Variations on Newton's telescope are still commonly in use. But most professional astronomical equipment follows a different design, invented around 1910 by Georg Ritchey and Henri Chrétien. To work properly, the parabolic mirror needs all the incoming rays to be parallel, which is fine if the object of study is, say, a single distant star. Today's astronomers tend to prefer wider fields of view. The trouble is that light entering from different angles can cause optical defects known as 'coma' in the image. Ritchey–Chrétien telescopes seek to combat these defects by using, instead of a parabolic mirror followed by a flat one, hyperbolic primary and secondary mirrors. This shape is less steeply curved than a paraboloid and better able to receive rays from broad angle, and the second hyperbolic mirror corrects distortion introduced by the first.

The Ritchey–Chrétien design is universally accepted to be the best telescope design so far. It is deployed, for example, in NASA's Hubble Space Telescope, which for over 20 years has been orbiting the earth at 17,000 miles per hour. Carrying a primary hyperbolic mirror measuring around 2.4 metres, it has captured over half-a-million images of some 30,000 objects in space. On earth, its images have spawned thousands of research papers and articles, and fuelled the latest thinking on questions such as the age of the universe and the possibility of life beyond our planet.

Awe-inspiring though this undoubtedly is, there is a catch in the Ritchey–Chrétien design. Hyperboloids are considerably harder to construct accurately than paraboloids, and even Hubble could not evade a glitch, discovered a few weeks after its launch in 1990. The primary mirror, it was found, was slightly the wrong shape, having been ground too flat at the edges by a tiny, but significant, 0.0022 millimetres. To the relief of astronomers worldwide, the error was successfully corrected in 1993. It is not just we frail humans who need to get our eyesight checked now and then and it is thanks to the geometry of optics that correctives can be applied.

THE BATTLE AGAINST DISEASE

Mathematical models and the spread of infection

The long history of humanity's attempts to conquer disease conjures up contrasting images – rough and ready folk-remedies, the 'cure-all' leeches and blood-letting relied upon over centuries, the once-revered and now discarded theories such as the balance of the body's four 'humours', right down to modern marvels such as the discovery of antibiotics and the intensive high-tech industry that is modern medical research. In today's world, mathematics too plays a notable role in the ongoing war against the spread of infectious disease. Modelling epidemics mathematically, and then investigating those models using computers, can provide genuine insight about what to expect once a real outbreak starts – invaluable information, when time is critical. It can help in planning a range of appropriate countermeasures, such as distributing treatments, launching vaccination programmes, or – in the case of non-human diseases – implementing culls or quarantines. These sorts of contingency plans can make profound differences, stopping epidemics in their tracks, and saving many lives; but they are also very costly, and not to be undertaken lightly if they are not truly necessary.

Answering to SIR

Modelling epidemics begins with what is known as the SIR model, originated by W.O. Kermack and A.G. McKendrick in 1927. It divides a population into three compartments: the Susceptible (those yet to be infected), the Infected, and the Recovered (who are assumed now to be immune). In this model, everyone must fit into one of these categories, so we ignore, for example, the fact that some people may have a natural immunity to the disease.

To make the mathematics easier, the SIR model makes several further simplifying assumptions. In particular, we suppose that that the size of the whole population is static, meaning that no-one is born and no-one dies. Needless to say, this is far from a realistic hypothesis, and might even be thought to be missing the point of a health crisis! Actually it is not so misguided; epidemics often move quickly enough that individual births and deaths are largely irrelevant to the numbers. Nevertheless, more sophisticated models can be developed with a dynamic population, which incorporates births and deaths. What is striking is that, despite its obvious limitations, the SIR model can capture several important aspects of an epidemic's behaviour.

If the population of the country is 1 million, then since, for SIR, this number is fixed, it must always be true that adding the three categories of person must total this amount, so the algebra is: $S + I + R = 1,000,000$. More generally, if p represents the size of the population, the first equation describing this model is that it must always be true that $S + I + R = p$.

Now, each of the numbers S, I and R will change, as the outbreak progresses, and it is the way that these numbers shrink or grow that is of interest. This necessitates understanding the

The basic SIR model.

laws governing how individuals transfer from one compartment to another.

The numbers emerging from the SIR model will determine the answers to crucial questions: will the disease take hold, or will the outbreak fizzle out? How quickly will it spread? How many people will be infected overall?

To get the answers to these important questions, we need to plug in two initial numbers. These will describe how easily people get sick, and how quickly they recover. Let's start with the rate of recovery, which we might call r. In mathematical terms, this is the rate at which people transfer from Infected to Recovered. The larger the value of r, the faster the rate of recovery, and so the shorter the time that people spend ill. In fact, we can be more precise: the average length of time that a person spends ill (i.e. in compartment I) can be expressed as: $\frac{1}{r}$

The value of r will vary from disease to disease, of course. Let's imagine a contagion, called *mathematitis*, which leaves people ill for seven days on average, so $r = \frac{1}{7}$. If one day there are 3,500 people ill (that is, in compartment I) then the following day we can expect one-seventh of them, 500, to recover, meaning that compartment R will grow by this amount.

Speed of recovery tells us nothing, however, about how easily people fall ill in the first place. This infection rate is measured by another number, which we can express as i. In the SIR model, this will determine how quickly people move from compartment S to compartment I. Some diseases are easy to catch (i.e. possess a high value of i), but are also quick to recover from (a high value of r). Others are less contagious (a small value of i), but will leave patients ill for longer (a low value of r). The most dangerous diseases are those that are both highly contagious and long-lasting – that is, which have a high value of i and a low value of r.

In analysing the infection rate, there is a subtlety that we did not see for recovery. The chance of a susceptible individual succumbing to the disease does not depend only the nature of the disease (specifically the value of i), but also on how many other infected individuals there are. Even in the case of a highly contagious disease, we will be safe so long as no-one near us has it. Equally, the spread of the disease depends not only on i, but on the number of susceptible individuals: if everyone has already been infected, then the disease cannot spread any further, no matter how contagious it is.

So the full definition of i is a little more complex: each sick person will infect a certain proportion of the susceptible population each day, and that proportion is i. That means that every ill person spreads the disease at a rate of $i \times S$ per day. So the overall rate of infection will be this multiplied by the number of people spreading the disease (namely I). Putting all this together, the number of people who transfer from S to I in one day can be expressed as the multiple $i \times S \times I$.

If, for our outbreak of *mathematitis*, we assume that $i = 0.001$, then that means each infected individual will infect one thousandth of the remaining susceptible population each day. If $S = 200,000$, then she will cause 200 people ($0.001 \times 200,000$) to become infected the next day. And if there are initially 50 people ill ($I = 50$) then the number increases correspondingly: $200 \times 50 = 10,000$.

Virulence and vaccination

In the SIR model, since we ignore births and deaths, the size of S can only decrease as people become infected, while that of R can only increase as people recover. The middle compartment I, meanwhile, can move either up or down as people enter it from S, and leave it for R. The curve that I traces out over time defines the character of the outbreak. To a great extent, this is determined by a single critical number, which describes the epidemic: the 'basic reproduction number', conventionally expressed as b. It defines the number of people who would be directly infected by a *single* unwell individual introduced into an otherwise susceptible population.

Experimental evidence has been gathered on the basic reproduction numbers of many diseases. For instance, that of smallpox is between 5 and 7, while that of measles is between 12 and 18.

In the SIR model, the value of b for a particular outbreak is defined as:

$$b = p \times \frac{i}{r}$$

where p is the country's population, i is the infection rate and r is the recovery rate. This number can tell us a great deal, in particular whether the outbreak is likely to take hold in the population, or whether it will fizzle out immediately. If the value is greater than 1 ($b > 1$), the outbreak will initially spread and grow. But if less than 1, $b < 1$, it will die out.

If we expand the model to take into account births and deaths, a new phenomenon appears: if the basic reproduction number is greater than, or equal to, 1, $b \geq 1$, the infection may become endemic, meaning that it will exist in the population at an approximately constant size, neither growing significantly in size nor dying out. Chickenpox is one such disease, endemic in almost every country in the world.

How many people will become infected over the course of the outbreak? Again, this will depend on the basic reproduction number. Specifically, in the SIR model, the proportion of population left uninfected by the end will be a number, x, which is between 0 and 1. Technically, x will satisfy the algebraic condition that $x = e^{b \times (x-1)}$, where e is the famous mathematical constant of around 2.718. Suffice to say that if *mathematitis* has a basic reproduction number of $b = 1.5$, then the value of x will be around 0.42, which translates into 42 per cent of the population uninfected by the time the disease plays out.

Fundamental to a community's public health is the level of vaccination against preventable diseases, such as mumps. The vaccinated not only guard themselves against becoming ill later, they prevent themselves from being carriers to others, so vaccination provides a means for disease-limitation through collective action. Mathematical modelling is clearly called for here, to gauge what level of immunization is required to prevent a disease from propagating through the population.

The immunization threshold (t) of a disease is the

proportion of people who need to be immune in order to prevent the infection from growing. It varies among diseases, but can be worked out from the basic reproduction number, b. Specifically, the two are related by the equation:

$$t = 1 - \frac{1}{b}$$

In the case of mumps, b is known to be between 4 and 7, so our equation would deliver an immunization threshold t of between 0.75 (that is, $1 - \frac{1}{4} = 0.75$) and approximately 0.86 (that is, $1 - \frac{1}{7} \approx 0.86$). This tells us that to be confident of keeping mumps at bay, public health officials and the medical establishment need to maintain an immunization level exceeding 86 per cent of the population.

Modifying the model

There are numerous ways in which epidemiologists have modified the SIR model to make it accord more with the real behaviour of infectious disease. The first obvious step is to incorporate births and deaths. Some diseases do not leave a legacy of immunity in those who recover, and so the model has also been adapted to allow for reinfection, whereby patients may again enter the Susceptible category: SIRS rather than SIR. Another variation splits the infection into a period of Latency, during which someone is contagious but showing no symptoms, followed by sickness: the SLIR model. Other factors that may be worked in include: individuals with natural immunity to the disease (which additionally may be assumed to be hereditary); countermeasures taken by the population, such as vaccination and treatments; some randomizing factor.

A more significant challenge is to model the epidemic spatially, taking into account the local geography. After all, diseases spread differently in rural and urban environments. An especially taxing problem for the modelling is to deal with the possibility that the pathogen will evolve a drug-resistant strain.

All these adaptations, and more, bring the SIR model closer to reality. But they may also make the resulting system of equations hugely more difficult to analyse, which is why the power of computing is increasingly important. When equations cannot be solved exactly, an arsenal of ingenious techniques of numerical analysis (see *Rain or shine?*) may be deployed to come up with approximate solutions.

These calculations are often computationally intensive. But it is through this type of mathematical and computational modelling, complementing conventional medical research, that the modern battle against disease is fought. Mathematics is not the same as treatment or prevention, but it can still be potently informative. As Lord Kelvin said, 'When you can measure what you are speaking about, and express it in numbers, you know something about it. But when you cannot – your knowledge is of a meagre and unsatisfactory kind.'

This is a serious matter – but not an entirely humourless one, as handled by a group of Canadian epidemiologists led by Philip Munz. In 2009 they brought light relief to the subject by analysing the epidemiology of Zombie outbreaks, as depicted in films such as George A. Romero's 1968 classic *The Night of the Living Dead*. Their SZR basic model had three states – Susceptible, Zombie and Removed. Susceptible people could become zombies through being bitten, or could die through natural causes.

Removed humans could then be resurrected as zombies, while zombies could in turn pass into the category of the Removed by having their heads severed or brains destroyed – the classic way to put paid to the walking dead. Munz and his team went on to analyse several variants of the model, incorporating well-known zombie-film conventions – such as a period of latency between being bitten and fully-fledged zombiedom. This produced the SIZR model, with a new Infected compartment I between S and Z. The results of their careful analysis would not surprise any horror-film devotee: even quarantine is unlikely to be enough to contain the outbreak; unless decisive action is taken at an early stage, a zombie epidemic will always tend in the long run towards the doomsday scenario, in which all humans are either zombies or dead.

This all goes to show the wide applicability of some mathematical techniques. When applied to matters of public health, they can help us grapple with very real matters of life and death. But the same ideas may be useful for shining a light on elements of popular culture.

WAVE-WORLDS

The mathematics of
sound and light

Talk about 'waves', and our immediate associations may be beach holidays, the sea lapping at our feet, or at least something watery. But, to a physicist, many phenomena consist in waves, light and sound being two of the most important. The understanding that sound is actually a wave, a matter of vibrations, goes all the way back to the Ancient Greeks. In the early 19th century it also became clear that light, too, is a wave, and at the same time the French mathematician and physicist Joseph Fourier (1768–1830) embarked on an intensive study of waveforms he studied heat transfer. His discoveries about the fundamental nature, and mathematical properties, of waves would have implications far beyond heat, and far beyond the intellectual climate of Napoleonic France.

Fourier's analysis began with the simplest waveform – the sine wave – together with some special relations that we recognize from music: the higher-pitched harmonics that accompany a note. Most importantly, Fourier understood how these basic waves could be combined to produce more complex waveforms, an insight that would go on to play a major role in the technological revolutions of the 20th century, from the synthesizing of music, to plumbing the oceans' depths, to investigating the stars.

The geometry of a wave

To visualize a sine wave, we might imagine a ball travelling around a circle, at a constant speed. If we ignore the ball's horizontal location and focus only on its vertical position, by plotting its movement up and down over time, we will produce a beautifully smooth and regular curve, known as a mathematician's sine wave (or sinusoid). Graphically, the wave appears when we trace out the ball's vertical position (y) against time (t) along the horizontal axis. It is described by the equation $y = \sin t$, or simply $y = S$. This sine wave is the most important example of the wave species, for in a real sense it is the source of every other.

It has a beautiful sister too, in the shape of a cosine wave, with equivalent equation, $y = \cos t$, or $y = C$. This is identical to the sine wave in every respect, but shifted backwards by a quarter of a cycle. In terms of the rotating ball, a cosine wave can be thought of as the ball's horizontal position.

The sine and cosine functions are familiar to generations of school pupils from a very different area: the geometry of triangles – or trigonometry. In this context, 'sine' is a relationship between the edges and angles of a right-angled triangle. Specifically, the sine of an angle is defined to be the length of the edge opposite (O) divided by that of the longest side (or hypotenuse H), giving:

$$S = \frac{O}{H}$$

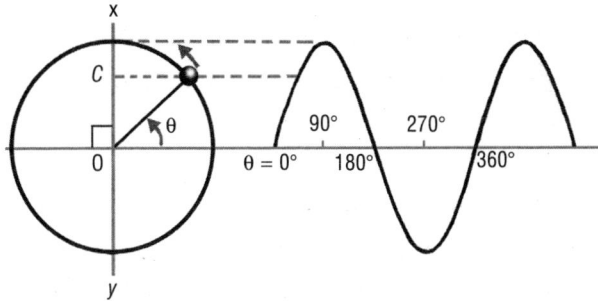

A sine wave can be seen in the vertical motion of a circular orbit, ignoring horizontal movement.

But what does this triangle have to do with the smooth wave we have just met? In fact, there is a triangle lurking in the picture of the rotating ball. We can see it if we imagine the ball attached to the centre by a rod. Now we have three lengths: the ball's horizontal distance from the centre, its vertical distance and the length of the rod. Since the horizontal and vertical distances are at right angles, they do indeed form a right-angled triangle, in which the rod is the hypotenuse (H), the vertical line is the opposite edge (O) and the third line is the horizontal distance.

To make things simple, we can assume that the length of the rod is exactly one unit, meaning $H = 1$. So, the sine of the angle is therefore $S = \frac{O}{1} = O$. This tallies perfectly with the fact that the sine wave appeared as a plot of the ball's vertical position.

Physical waves are usually described by two properties: their frequency (normally measured in hertz) and their amplitude (measured in metres). The number of complete cycles a wave travels through in one second defines its frequency; a sound of 1 hertz corresponds to one complete cycle per second, while Middle C has a frequency of around 261.6 hertz. Higher-frequency soundwaves produce higher pitches, while lightwaves of different frequencies are interpreted by our brains as different colours. Green light, for instance has a frequency of around 540,000,000,000,000 hertz (or 540 terahertz). Closely related to the frequency, the wavelength is the distance between peaks of the wave. For middle C this is around 1.3 metres, while for green light it is around 0.0000006 metres, or 600 nanometres. (For any wave, doubling the frequency means halving the wavelength, and vice versa.)

If frequency and wavelength concern the spaces between peaks of the wave, they say nothing about how big those peaks are. The second basic measurement of a wave is its amplitude. For the orbiting ball, the amplitude was given by the distance from the ball to the centre of the circle, which we set to 1. With sound- or lightwaves a larger amplitude means more energy being carried by the wave. Our brains interpret amplitudes as volume in soundwaves, and as brightness in lightwaves.

These qualities can all be captured in mathematical notation. If we start with a sine wave with a frequency of 1, which is to say $y = \sin t$, then a wave with twice the frequency will have an equation $y = \sin 2t$ (which corresponds to the ball moving round at double the speed). Amplitude, meanwhile, can be adjusted by scaling up the whole wave, by doubling the length of the rod, expressed by the equation $y = 2 \sin t$. For soundwaves, this produces a wave twice as loud.

Certain levels of increase in the frequency of a soundwave have special significance – recognized instinctively by most people. Doubling the wave's frequency creates a sound that somehow sounds the same, but at a higher pitch: it has jumped to what musicians a call the octave above, but what mathematicians call the 'second harmonic' (the first being the starting note). Harmonics can be produced on stringed instruments by touching the active string in the centre to keep it stationary – this has the effect of halving the wavelength, thereby doubling the frequency. The second harmonic of the sine wave has the equation $y = \sin 2t$, as stated above, which we can also denote as S_2, for short. Higher harmonics come from increasing the frequency by other amounts or equivalently reducing the wavelength. The third harmonic, for example, has a frequency three times that of the root note. For the sine wave, this is expressed as $y = \sin 3t$, or simply S_3. In musical terms its sound is an octave plus a fifth higher than the base note.

With a little practice, as musicians discover, it is not too difficult to understand the effects of varying a wave's frequency and amplitude. But that's not all there is to the difference in the quality of a sound; after all, we would hardly mistake a flute and a violin for one another, even playing the same note at the same volume. There is something else at work, which musicians refer to as 'timbre'. This phenomenon gives music an extra layer of richness, and derives from the fact that physical waves rarely follow the simple pattern of the mathematicians' sine wave. With the right equipment, we can see this directly. Viewed on an oscilloscope, musical instruments' waves typically have far more complicated appearances than the sine wave, and differ from each other in subtle ways.

Nevertheless, all waveforms have the same essential structure – a single pattern, which repeats itself over and over again. And it is this regularity that provides our route to understanding the geometry of waves.

Interesting interference

Thus far we have considered waves in splendid isolation, in an uncluttered world. But in reality, waves are not sealed off – rather, they interfere with one another. This is most clearly illustrated by our starting point, water waves. Scientists are able to investigate wave behaviour using 'ripple tanks', where they can experiment with different frequencies and amplitudes. Set one paddle vibrating in the tank, and water will ripple outwards in circles; but with two paddles we enter the realms of wave interference. At places where the two waves are in sync they will reinforce each other, producing a wave of double the amplitude; but in other locations where they are exactly out of sync, they will cancel each other out leaving flat water. Interference may seem a complicated process, but in mathematical terms it is nothing more than addition. If V and W are two waves, then their combined waveform, where the two interfere, is given by the sum $V + W$.

There is nothing special about water here, for the same phenomenon applies to other types of wave too, including sound. Indeed, this understanding lies behind the technology of today's noise-cancelling headphones. If the ambient sound is described by a wave W, then the headphones will try to produce the exact opposite: $-W$. When these two combine, the idea is that they will cancel out precisely, leaving the wearer in total silence: $W - W = 0$.

So while mathematicians enjoy the elegant symmetry of the sine wave, the wider world – and human creativity – has many messier waveforms to offer. How can noise-cancelling headphones produce exactly the right wave? The same question recurs in many guises, for instance in the context of music

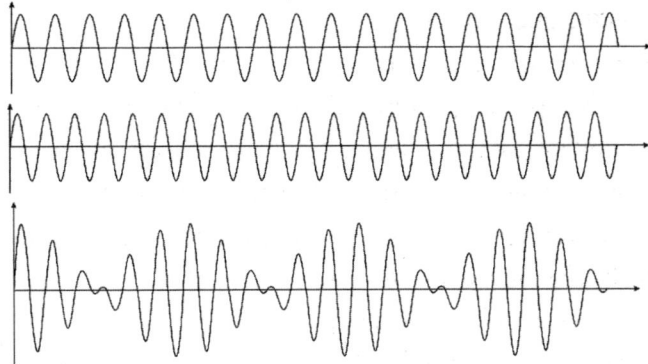

Complex waveforms appear when two waves interfere, reinforcing each other at some points, and cancelling each other out at others.

sampling. How does a computer *recreate* the sound of an instrument such as a trombone?

The answer in every case is provided by the branch of mathematics known as 'harmonic analysis', which includes techniques for building up complex waveforms by adding sine waves together, in such a way that that they interfere in just the *right* manner.

The centrepiece of the theory is Fourier's Theorem, named after the eminent Frenchman of Napoleonic times. This astonishing theorem asserts that almost any waveform can be constructed by starting with a sine wave, and then adding its corresponding cosine wave together with their various harmonics, each adjusted to a suitable volume.

In mathematical terms, say we start with some waveform W, perhaps representing the sound of a trombone, which we wish to recreate artificially. The easiest wave for a computer to produce is the sine wave, S (or S_1), and it is not hard to adjust this so that its frequency matches that of W. That is the starting point, and

THE SAWTOOTH WAVE: HARMONIC ANALYSIS AT WORK

While the sine wave is beautifully smooth and symmetrical, many other waveforms are not. For instance, the 'sawtooth wave' is, as its name suggests, covered with sharp corners, so provides a good example of Fourier's method and harmonic analysis at work. Despite appearances, it can indeed be expressed using a sine wave and a suitable combination of its harmonics. In some ways, this example is more straightforward than others, since it does not require the cosine components, while the volumes for the harmonics turn out to be quite simple.

A sawtooth wave

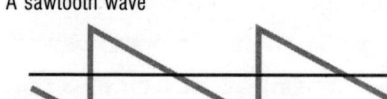

An approximation up to S_4

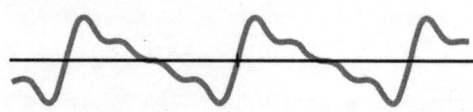

An approximation up to S_{20}

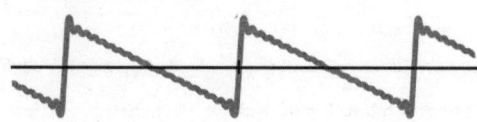

The starting point, as usual, is a sine wave with the same frequency as the sawtooth, which we call S_1. Next, take its second harmonic, S_2. It turns out, using Fourier's work, that this should be played at a volume half that of the original:

$$\frac{1}{2}S_2$$

When the original sine wave and this harmonic interfere, it produces the combined wave:

$$S_1 + \frac{1}{2}S_2$$

On top of this, the third harmonic is played at a volume of one third, and added on too:

$$S_1 + \frac{1}{2}S_2 + \frac{1}{3}S_3$$

Continuing the obvious pattern will produce a sequence of ever better approximations to the sawtooth wave, with the exact formula as the infinite limit:

$$W = S_1 + \frac{1}{2}S_2 + \frac{1}{3}S_3 + \frac{1}{4}S_4 + \frac{1}{5}S_5 + \cdots$$

This can be summarized as:

$$W = \sum \frac{1}{n}S_n$$

And it is a mathematical description of the angular sawtooth wave purely in terms of smooth sine waves.

we will also need the corresponding cosine wave C (or C_1).

Each of these waves comes with a whole family of harmonics, which it is a simple matter for a computer to produce: S_2, S_3, S_4, etc. along with C_2, C_3, C_4, etc. What Fourier tells us is that there are some numbers a_1, a_2, a_3, ... and b_1, b_2, b_3, ... that represent the volumes to which these sine and cosine waves need to be set. Making these adjustments creates a sequence of sine waves $a_1 S_1$, $a_2 S_2$, $a_3 S_3$, ... and of cosine waves $b_1 C_1$, $b_2 C_2$, $b_3 C_3$, ... The magic occurs when *all* of the resulting waves are allowed to interfere. If we have done the job properly, they will combine to create exactly the wave we want to synthesize. This may be written out as:

$$W = a_1 S_1 + b_1 C_1 + a_2 S_2 + b_2 C_2 + a_3 S_3 + b_3 C_3 + \cdots$$

This lengthy equation can be written more concisely using the Sigma notation for adding things together (with the conventional subscript n denoting the number that changes):

$$W = \sum a_n S_n + b_n C_n$$

It is no surprise that a large collection of varied waveforms can be constructed using this method. But it is very remarkable indeed – and extraordinarily useful – that *all* can. However, there is one missing step: to successfully synthesize the sound of a trombone, we need to know the volumes a_n and b_n to which the various harmonics need to be set. Fourier completed his work by providing a method for discovering exactly these values, describing a subtle piece of mathematical analysis to be carried out on W, which will tell us the appropriate volumes to play the various harmonics.

Following Fourier

Joseph Fourier's work on waveforms, encapsulated by his wonderful theorem, has been of incalculable value to the modern world. Most obviously, it is central to all sorts of acoustic technology such as sound-sampling, CD-recording and digital radio, in which sounds are stored and reproduced as a series of interfering sine waves, each tuned to the correct volume, precisely according to Fourier's method. But its influence runs much further than this, and today scientists in a range of disciplines appeal to Fourier's ideas, even where no waves are obvious. For instance, some methods of cracking codes or multiplying large numbers quickly make essential use of his insights.

Scientists even rely on Fourier when probing the nature of matter. The internal structure of a crystal, for example, can be determined by the way lightwaves scatter off it; since waves of different frequencies will scatter in different ways, it is only with Fourier's help that researchers can make sense of the resulting patterns. Indeed, whether you are an oceanographer using sonar to investigate the Atlantic's depths, an astronomer deploying a radio-telescope to scan the night sky – or simply listening to music while you surf the web using Wi-Fi technology – you are quietly communing with Fourier.

SEARCH-ENGINE SOCIETY

The algorithms behind
Google's PageRank

Science fiction has long enjoyed the idea of travellers jumping back and forth between parallel worlds, in which two subtly different versions of life go on. Some would say that with the rise of the internet since the 1980s we have all been leading this double life, constantly skipping back and forth between our physical and virtual existence. Our high-street shopping is mirrored by the experience of clicking over to an online retailer; we may share holiday photos with our friends over coffee in a café, or via Facebook; we can whisper our observations on life to a colleague or broadcast them to the world on Twitter; and for longer discussion, or simply letting off steam, there is the Blogosphere. And so it goes on. Even if one chooses not to immerse oneself in social networks, the World Wide Web has taken over as our first port of call for finding out about almost anything, whether that be tomorrow's weather or our ancestors, supermarket produce or secret pleasures.

The Web now constitutes an extraordinary mine of information, embracing several trillion – several million million – interlinked webpages. But without a system of navigation, a way to connect the searcher with the 'searched for', the technology – even at a tiny fraction of its current size – would be all but useless. It is therefore little surprise that a new verb, 'to google', has entered the language in recent years, a testament to the appeal and market dominance of one particular company, but more importantly an affirmation of the essential contribution of the internet's compass: the search engine. Naturally, it is knit together by mathematics, and its threads are worth unravelling.

Crawling and caching

At the headquarters of Google and other search engines, maps of the internet are constantly being built and updated. Such a map contains cached versions of all the webpages that have been explored – copies stored in a stripped-back format. The job of building and maintaining this map is undertaken by an army of 'web-crawlers', who forage in waves, with a crawler being sent to explore a section of the Web based on the results of previous searches; in turn it supplements the map with any new links or pages it finds, marking them out to be thoroughly searched in the next wave. Today's search engines are refreshing and updating their internet maps several times per day.

Once the crawlers and cachers have done their jobs, the next phase involves the indexing function, which compiles an index of the words and phrases appearing on all the cached pages. For a search topic – say Berry's Paradox (discussed in *The perils of paradox*) – the index will list the cached pages containing the text, along with their addresses on the Web, much as the index at the back of a book directs readers to relevant pages.

Internet caches and indexes are undeniably technological

marvels, but alone they do not solve the problem of internet searching. Simply presenting the user with the entire index entry for Berry's Paradox is unlikely to satisfy them; after all, trillions of webpages are built from the same collection of 10,000 or so of the most common words. At the time of writing, Berry's Paradox returns over 3 million search results, the overwhelming majority of which are likely to be of minimal use to the would-be researcher.

Evidently a further stage is needed, to sort results into an order likely to be useful. The most famous technique for doing this is Google's PageRank algorithm, designed by the co-founders of Google, Larry Page and Sergey Brin. The idea is to rank all the pages on the Web in order of influence, and the mathematics behind it relies on two techniques: network theory and matrices. What follows is a simplified account of how it works.

PageRank principles

The first step is to model the internet network mathematically. We can begin by representing each webpage as a node on a piece of paper; to make this easier, let's assume we are dealing with a reduced version of the Web containing just four pages – A, B, C and D – some of which are linked to each other. The idea is to connect two nodes with an edge (line) if the corresponding webpages are linked.

Of course, weblinks are not symmetric: it may be that page B links to page A, but A does not link to B. So instead of plain edges, we use arrows to suggest the direction of the linkage creating what mathematicians call a directed network. For these two pages, there are four possibilities: neither A nor B link to each other, A links to B but not vice versa, B links to A but not vice versa, and each of A and B links to the other. With the network laid out, we next need to pull this information into a more usable form, which is where we have recourse to the algebra of matrices.

A matrix is simply an array of numbers, organized into columns and rows. The most useful come in the form of a square, such as a 3×3 matrix:

$$\begin{pmatrix} 1 & 2 & 3 \\ 4 & 5 & 6 \\ 7 & 8 & 9 \end{pmatrix}$$

A vector, meanwhile, looks like a single column of a matrix, perhaps

$$\begin{pmatrix} 0.1 \\ 0.2 \\ 0.3 \end{pmatrix}$$

From the network of four nodes we jotted down, a matrix can be extracted encapsulating the relationships among the various pages. If we take just one webpage, A, its outgoing links may be expressed as a vector:

$$\begin{pmatrix} 0 \\ 0 \\ 1 \\ 1 \end{pmatrix}$$

The top '0' states that page A does not have an outgoing link to itself – we will always make this presumption, since it doesn't make sense for a page to bolster its own rankings by linking to itself. The '0' in the second row tells us that page A does not link to page B either, but the final two 1s show that it does link to C and D.

We can similarly compile the vectors for the outgoing links of B, C and D. Putting them side by side produces the 'adjacency matrix':

$$\begin{pmatrix} 0 & 1 & 1 & 1 \\ 0 & 0 & 0 & 1 \\ 1 & 0 & 0 & 1 \\ 1 & 0 & 0 & 0 \end{pmatrix}$$

This captures everything we need to know about the network, and a useful new feature emerges: when read vertically, the columns represent each page's outgoing links, but when read horizontally the rows correspond to its incoming links. So we can see from the top row that page A has links from each of B, C and D, while page B has only one incoming link, from D, and so on. This adjacency matrix does the groundwork for extracting the all-important page rankings – but we need to delve a little deeper into the mathematics of matrices to continue the PageRank story.

Many scientific processes can be represented as a vector multiplied by a matrix, and the process is equally important for webpage rankings. Suppose we take a vector, $\begin{pmatrix} 0.1 \\ 0.2 \\ 0.3 \end{pmatrix}$, and we wish to multiply it by a single row of a matrix, say (1 2 3). All that is required is to multiply together the individual entries, and then add them up

$$(1\ 2\ 3) \begin{pmatrix} 0.1 \\ 0.2 \\ 0.3 \end{pmatrix} = 1 \times 0.1 + 2 \times 0.2 + 3 \times 0.3 = 1.4$$

There is one requirement for this process to work – that the width of the row be the same as the height of the column. To

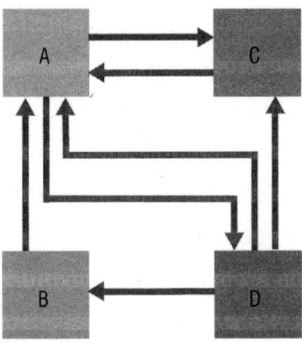

A network representing a four-page web.

multiply the vector by a larger matrix, say:

$$\begin{pmatrix} 1 & 2 & 3 \\ 4 & 5 & 6 \\ 7 & 8 & 9 \end{pmatrix}$$

we simply proceed row by row. Taking into account the 1.4 we've already calculated, that means:

$$\begin{pmatrix} 1 & 2 & 3 \\ 4 & 5 & 6 \\ 7 & 8 & 9 \end{pmatrix} \begin{pmatrix} 0.1 \\ 0.2 \\ 0.3 \end{pmatrix} = \begin{pmatrix} 1.4 \\ 4 \times 0.1 + 5 \times 0.2 + 6 \times 0.3 \\ 7 \times 0.1 + 8 \times 0.2 + 9 \times 0.3 \end{pmatrix} = \begin{pmatrix} 1.4 \\ 3.2 \\ 5.0 \end{pmatrix}$$

The purpose of the PageRank algorithm is to assign every webpage a measure of its influence. The essential idea is that pages with a greater number of incoming links should receive a higher rank than those with fewer, since they are more influential. But this alone is not enough. Imagine someone wishing to promote their own website sets up thousands of dummy pages linking to their own. Then, hey presto, these thousands of incoming links would give them a hugely inflated rank. Clearly this is not acceptable, so instead we only really want to count links from pages that themselves have some worth.

What is needed is a measure of a page's influence that is inherited by the pages it links to: incoming links from popular pages should be worth more. With some clever mathematics, this idea can be made precise. Suppose our webpage has influence of 20 by some measure, and it in turn links to 5 other pages. Each of these other pages should receive an extra $20 \div 5 = 4$ units of influence from us. The total influence of those pages can then be worked out by adding up the influence they receive from each incoming link.

It is an elegant idea and it certainly solves the dummy pages problem, since they will each have negligible influence. But on the face of it there is a difficulty: to calculate the influence of a page, it seems we first need to know the influence of all pages linking to it. Where does the chain of calculations begin? Of course, it is impossible to get to a beginning – the Web has no 'first page', and furthermore, there are likely to be loops. If page A links to page B, which links to page C, which in turn links back to A, then how can we begin to calculate their influence? This seems a serious obstacle, yet the mathematics of matrices can provide a way through the maze.

Out of the matrix maze

In our experiment above, we divided the influence of our webpage (20) by the number outgoing links (5). We can use this principle to amend the adjacency matrix. By totalling the figures in our original vector for page A – 0,0,1,1 – we find we have two outgoing links. Our next manoeuvre is to divide every entry of the vector by this figure 2, giving:

$$\begin{pmatrix} 0 \\ 0 \\ \frac{1}{2} \\ \frac{1}{2} \end{pmatrix}$$

The point of that move is this: if our webpage has only one outgoing link, that link will contribute more to the linked page than if our webpage indiscriminately links to thousands of others. Dividing by the total number of outgoing links in this way expresses the degree to which our page's influence is shared among them.

Having done this for page A, we can do the same for pages B, C and D and assemble the resulting vectors into a modified adjacency matrix, which we might call M:

$$M = \begin{pmatrix} 0 & 1 & 1 & \frac{1}{3} \\ 0 & 0 & 0 & \frac{1}{3} \\ \frac{1}{2} & 0 & 0 & \frac{1}{3} \\ \frac{1}{2} & 0 & 0 & 0 \end{pmatrix}.$$

Remembering that our ultimate goal is to assign each page a measure of its influence, its rank, we can denote that measure for our four pages by r_A, r_B, r_C and r_D. It is handy to assemble these into vector form:

$$r = \begin{pmatrix} r_A \\ r_B \\ r_C \\ r_D \end{pmatrix}.$$

For the ranking system to work as required, it must be that the page A inherits influence from its incoming links B, C and D (remembering that these are listed on the top row of the matrix M.) This gives us an equation that the numbers should satisfy:

$$r_A = r_B + r_C + \frac{1}{3}r_D.$$

The other rows of matrix M produce similar equations for pages B, C and D; but what is the exact connection between these four equations and the matrix M? In fact, it is much simpler than it looks, being nothing more than the matrix multiplication we met earlier. The four equations can be combined into a single, simple equation relating the matrix M to the vector r:

$$M \times r = r$$

This equation expresses the fundamental condition that the page-rank vector r must satisfy. In the jargon, this says that r is a 'stationary vector' of the matrix M – in other words, multiplying by M leaves r unchanged.

Our goal has now become clearer: we need to calculate the stationary vector r, and that will reveal the influence of each of the four webpages. In fact, mathematicians have been doing similar calculations since the 18th century and have assembled a battery of standard algebraic techniques that can be brought to bear on the problem. However, when real search engines perform this calculation they face an additional complication: of course the number of pages involved is rather more than four. Instead, they have to address a matrix of size $n \times n$, where n is several trillion. In such a situation mathematicians' favourite tricks become impracticably slow.

There is, though, one secret weapon – a so-called 'power

method' – that can be deployed to find the stationary vector. The procedure was invented in 1913 by Herman Müntz. Its general principle is to begin with any vector whose entries add up to 1, which we might call t. For example:

$$t = \begin{pmatrix} 1 \\ 0 \\ \vdots \\ 0 \end{pmatrix}$$

and then repeatedly multiply it by the matrix M. It is a mathematical fact, given certain technical assumptions, that the sequence of vectors $t, Mt, M^2t, M^3t, \ldots$ will eventually converge on the true stationary vector, which in turn will tell us the rank of every webpage. For practical purposes, it is usually enough to terminate the calculation at around $M^{100}t$.

As well as being calculable in a practical way, Müntz's method has another advantage: it ties in with another intuitive way to understand the rankings of webpages, in terms of the behaviour of a random internet surfer.

The rovings of the random surfer

Imagine – and it might not take too much imagining – filling some minutes of boredom with some mindless web-surfing, clicking on links and moving from page to page absolutely at random. We are more likely to find ourselves moving towards the more highly influential pages, widely linked to from across the Web, than obscure pages with few incoming links. In fact, we can quantify this probability more precisely: after a long enough phase of random clicking, the likelihood that we end up on a particular page will be quantified exactly by its page-rank.

We can illustrate the principle by returning to our mini-web of four pages, starting with our vector t:

$$t = \begin{pmatrix} 1 \\ 0 \\ 0 \\ 0 \end{pmatrix}$$

which we can interpret as saying that we begin randomly surfing at webpage A. If we multiply this vector by matrix M we arrive at

$$Mt = \begin{pmatrix} 0 & 1 & 1 & \frac{1}{3} \\ 0 & 0 & 0 & \frac{1}{3} \\ \frac{1}{2} & 0 & 0 & \frac{1}{3} \\ \frac{1}{2} & 0 & 0 & 0 \end{pmatrix} \begin{pmatrix} 1 \\ 0 \\ 0 \\ 0 \end{pmatrix} = \begin{pmatrix} 0 \\ 0 \\ \frac{1}{2} \\ \frac{1}{2} \end{pmatrix}$$

This result can be read as telling us that after *one* click, we have a 50 per cent chance of being on each of pages C and D. Multiplying this vector by M again would give us:

$$M^2t = \begin{pmatrix} \frac{2}{3} \\ \frac{1}{6} \\ \frac{1}{6} \\ 0 \end{pmatrix}$$

which tells us the probabilities of the surfer's location after *two* clicks, while after *three* clicks the vector is:

$$M^3t = \begin{pmatrix} \frac{1}{3} \\ 0 \\ \frac{1}{3} \\ \frac{1}{3} \end{pmatrix}.$$

No pattern is yet obvious in this sequence; but after eight or so iterations, the sequence of vectors ceases to change quite so dramatically from one click to the next. In fact, relatively quickly it settles on a final value for r:

$$\begin{pmatrix} \frac{3}{7} \\ \frac{1}{14} \\ \frac{2}{7} \\ \frac{3}{14} \end{pmatrix}$$

This is the goal we have been seeking – the vector that expresses the relative influence of our four webpages. In the long run, if we surf our mini-Web at random, then at any given moment we will have a $\frac{3}{7}$ chance of being on page A, a $\frac{1}{14}$ probability of finding ourselves on page B, and so on. It is also a simple matter to check that this vector satisfies the fundamental equation $Mr = r$. As expected, in our mini-web A is the most influential page, followed by C, D and finally B.

Following the money

'Basically, our goal is to organize the world's information and to make it universally accessible and useful. That's our mission.' These were the words of Google's Larry Page in 1998. Since that time, a whole family of variations, descendants and add-ons have more than kept up with the internet's rapid expansion, making it a more easily navigable place. The issues are not merely technical of course. Ultimately, the true value of a webpage to its viewer is subjective, so search engines are general signposts – as if the librarian is sending you to the right bookcase, but not picking out a particular title for you.

Nevertheless, the race is on to understand more and more about where people are going, why they are going there, and where they might go next. This raises new issues. After all, Google and its competitors are not charitable enterprises, and the desire for accurate page-ranking and keyword tracking data is becoming ever more tightly tied to the charging models for online advertisers and the ever-expanding reach of internet commerce. Do people accept a trade-off between ease of use and having their online activity scrutinized by companies keen to sell them products and services? At the moment the answer seems to be 'yes', such is the boon of internet access. Yet this debate is destined to be reopened time and again. While mathematical advances can provide us with algorithms with the power to change our online lives, decisions about whether and how to employ them must ultimately be taken in the social and political spheres.

HOLD THE LINE, PLEASE!

The mathematics
of queues

By tradition, the British are addicted to it. Some ill-mannered nationalities – if you believe the stereotype – cheat at it. Few but the most powerful can avoid it. But *some* have rather enjoyed theorizing about it, with profoundly significant consequences.

Queuing, or waiting in line, is a tiresome activity, but simply a fact of life in a crowded society with any semblance of organization. So integral is it to modern life that we might well be surprised to acknowledge the many varieties of queuing that fill our days. To the obvious and familiar manifestations – pausing at traffic signals, standing at the supermarket till, waiting for a bus, the annoyance of call centres – the Information Age and telecommunications revolution have added additional layers of electronic queues that we are subject to, whether we know it or not.

The need to plan efficient systems for organizing people, procedures and information has given rise to the subtle mathematics of queuing theory, whose history goes back to some innovative Scandinavian analysis of early telephone exchanges. Today this knowledge informs an array of practical applications, from the design of parallel processing systems in computers to improving the efficiency of factories.

Traffic harmonics

In any sensible shop, where the management does not want a horde of complaints, the order of serving customers seems self-evident. It is the model of 'first in, first out', where the customer who has been waiting the longest – and reaches the front of the queue – gets served next. But when the queue is not of people, but of insentient objects unlikely to feel aggrieved by queue jumping, other models are possible.

Someone washing dishes in a restaurant may, as the waiters pile up the dirty plates, find himself constantly taking from the top of the pile, in which case the latest dishes to arrive get washed before the earlier ones, which sit nearer the bottom of the pile. He would be adopting the 'last in, first out' system, also known as a stack. So the very last dish to be washed will be the one that has been waiting longest.

In computer science, stacks have been common since the 1950s. They naturally emerge as the order in which processes are opened and closed. A programme (call it A) may invoke a procedure B, which might in turn call up a sub-routine C. So A is the first task to be begun, but then B is placed on the stack followed by C, but then in shutting the sequence they need to finish in reverse order, with C first followed by B and finally A.

Of course, an ideal orderly queue would see joiners arrive at the same even rate as others depart. But in life things are not so simple, and nowhere is the tendency towards 'bunching' experienced with such irritation than in road traffic. This is inevitable

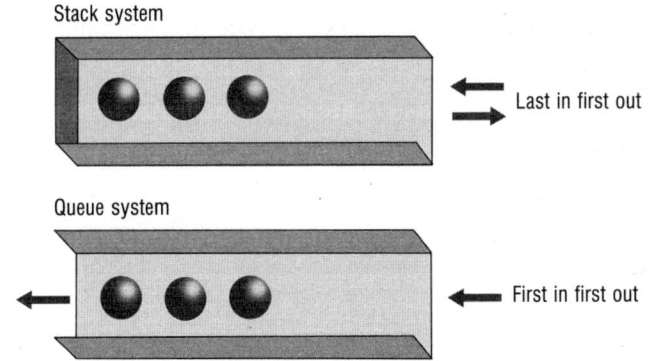

Stack systems are more common in computer science than
traditional queues

if, for example, a group of cars drive along a road where overtaking is impossible. But how *many* bunches will there be? The answer is surprisingly elegant. To begin with, the first car will be able to travel at whatever speed it likes, so far as road conditions permit. For every subsequent car, its own speed will be limited by the one in front. If a driver wishes to go faster than the one ahead, then she has no choice but to bunch up behind it. But if she is happy to go slower, then she can travel at her preferred speed, in effect starting a new bunch with herself at the head. The key insight is that each car begins a new bunch if it is the *slowest so far*.

Thus the question about 'how many bunches' translates into 'how many cars are likely to be the slowest so far?' The first car will certainly fit that description, and is guaranteed to start a bunch. The second then has a 50 per cent chance of being slower; if it is, it will also begin a new bunch, taking the total to 2; if not, it will remain with the initial bunch, keeping the total number of bunches at 1. So, the average number of bunches in two-car queue is $1\frac{1}{2}$. On the same principles, the third car has a $\frac{1}{3}$ chance of being the slowest so far, and the average number of bunches after three cars will be:

$$1 + \frac{1}{2} + \frac{1}{3}$$

As this process continues, the average number of bunches in a line of n cars will this be:

$$1 + \frac{1}{2} + \frac{1}{3} + \cdots + \frac{1}{n}$$

This is, in fact, the famous mathematical object known as the harmonic series, and a classical result has a bearing here. A natural question – mathematically if not in terms of traffic management – is what happens in the long run, as more and more cars enter the road? Will the average number of bunches increase without bound or eventually settle down to some limiting value? As the 14th-century philosopher Nicole Oresme showed, although this sort of series grows extremely slowly (it becomes increasingly unlikely that the next car will be the slowest so far), it does nevertheless grow without bound: leave it long enough and the number of bunches will eventually surpass any number at all. But to reach an average of just ten bunches would require over 12,000 cars on the road!

Factorial fun

Clearly, for a traffic queue on our 'no overtaking' road, there's not much that individual cars can do to alter their position in the queue. But the situation may be quite different in other scenarios. Let us suppose five friends (Adam, Barbara, Clara, David and Edward – or A, B, C, D and E) are sitting in a park when an ice-cream van arrives. If they all run over to buy ice-cream, there are several possibilities for the order the queue might take – but how many? The answer is given by the mathematician's factorial function. There are five possible choices for the person in front – any of A, B, C, D and E; with that position occupied, there are then four choices for second place. This means that the number of ways of filling the first two spots is $5 \times 4 = 20$. Continuing this line of thought, the total number of ways of arranging the five people in order is $5 \times 4 \times 3 \times 2 \times 1 = 120$. This is known as 'five factorial', and is usually denoted with an exclamation mark, so in this case: $5! = 120$.

Factorials are central to the analysis of queues, and they can be invoked to answer more complex questions. If, there are eight friends chatting, rather than five, and just three of them are sent to join the queue (perhaps buying ice-creams for the rest), how many configurations are possible for deciding who is to go up and in what order? Now there are eight possibilities for the person at the head of the queue, then seven possibilities for the next in line, and six for the third, giving an answer of $8 \times 7 \times 6 = 336$ possible queues.

In term of factorials, it is as if we start calculating $8!$ but then give up when we get to 6. Or, to put it another way, it is $8!$ with $5!$ removed $\frac{8!}{5!}$.

This does indeed deliver 336 possibilities. The reason for the appearance of 5 here is that it is 8 (the total number of people involved) minus 3 (the number of queuers). The general rule here is that the number of permutations (i.e. different queues) of r people from a total group of n is:

$$\frac{n!}{(n-r)!}$$

In our example, $n = 8$ and $r = 5$.

There is one other variation on the theme that we might wish to analyse: if the order *within* a group of three friends does not matter – say they are not fussy about who's at the head of the ice-cream queue – we might yet want to know how many differently constituted groups of three might head to the van. With this calculation, we only define two queues as distinct if they do not contain exactly the same people. Otherwise they are the same, regardless of order. We know there are 336 total possible queues of three; but we also know that each choice of three people can be ordered in $3! = 6$ different ways. So the answer here will be:

$$\frac{336}{6} = 56$$

different possible *combinations* of three people from the total group of eight.

In general, the number of combinations of r people from n is the number of permutations, using the formula above, divided by the number of ways of ordering a single queue of r people, which, using the factorial notation, is $r!$ This produces an overall formula:

$$\frac{n!}{r! \times (n - r)!}$$

'Your call is important to us'

While we are all familiar with the kind of physical queuing represented by traffic and waiting for ice-creams, there are other contexts where we may well be unaware of our fellow queuers, as happens whenever we dial an organization's call centre. If anecdotal evidence is anything to go by, call centres have their work cut out in managing their queues more effectively. But how many workers does a call centre actually need? The dimensions of this sort of question make it one of surprising mathematical depth and importance, well beyond the customer-services industry.

A naive attempt to answer it might go as follows: if the call centre receives on average 8,000 calls per 8 hour working day, then that amounts to 1,000 calls per hour. If each call is on average 2 minutes and 24 seconds long, then each worker can manage 25 calls per hour. So the number of phone operators needed at each moment is $1,000 \div 25 = 40$. For all its appealing simplicity, this type of thinking is likely to lead to disaster, since calls are unlikely to arrive in regular sequence at intervals of 2 minutes 24 seconds. There will be gluts and there will be gaps.

A more sophisticated approach requires the mathematics of queuing theory, and begins by measuring the call centre's ability to answer calls with a single number, P. This represents the probability that when someone calls up, they are unable to get through. (For purposes of our discussion, we assume call-waiting is not available – the caller will either get through or not.) The higher the value of P, the worse the call centre is doing.

The manager of the call centre will closely monitor the value of P, which is easily defined: it is the number of incoming calls that do not reach a customer service adviser, divided by the total number of incoming calls that day, whether they get through or not. We might write this as:

$$P = \frac{\text{Failures}}{\text{Total Calls}}$$

Now, the manager may want to impose a target here, declaring that no more than 5 per cent of callers should find themselves unable to get through, making a target of $P = 0.05$. Well, this is easily said, but what we really need to know is how many phone operators are needed to achieve this target. And that means we need two pieces of information: how many incoming calls there are, and how long each conversation takes.

Conveniently, there is a single measure of telephone traffic that automatically takes both factors into account. It answers to the question 'If all calls were to get through, how many phone-lines would be in use for what proportion of the time?' There is even a proper name for a unit of telephone traffic – the Erlang, named after Agner Krarup Erlang, the Danish telephone engineer whose analysis in the early 20th century kick-started the subject of queuing theory. One Erlang is defined to be the volume of telephone traffic that keeps exactly one telephone line in constant use. Equivalently, one Erlang will see two telephone lines each used half of the time, or three lines used a third of the time, and so on.

The manager, we remember, will need to know the proportion of callers who currently cannot get through to a service adviser – the critical number P. In 1917 Erlang devised a formula, known as Erlang-B, to provide an expression for P: it became the central formula in queuing theory. It depends on two things: the volume of incoming telephone traffic (E, measured in Erlangs), and the number of phone operators (n) at the call centre.

If we assume an average of 5 Erlang of traffic (so $E = 5$, which means enough incoming traffic to keep five telephone lines in constant use, or ten in use half of the time, etc.), and the call centre is staffed by 4 people ($n = 4$), the value of P is expressed as a fraction. The *top* part of it is the expression $\frac{E^n}{n!}$ (where the exclamation mark represents a factorial, as above). So, using the figures of our example, this first ingredient will be equal to:

$$\frac{5^4}{4!} = \frac{5 \times 5 \times 5 \times 5}{4 \times 3 \times 2 \times 1} = \frac{625}{24}$$

which comes out at around 26.042. The second ingredient, which goes on the *bottom* of the fraction, is several similar terms added together. More specifically, it is the number:

$$1 + E + \frac{E^2}{2!} + \frac{E^3}{3!} + \cdots + \frac{E^n}{n!}$$

In other words, the power of E increases by 1 from left to right, as does the factorial number up to n (the number of staff). Why does this begin with $1 + E$?. Well, that is simply a more straight-forward way of presenting the first two terms in the sequence, which presented more formally would be:

$$\frac{E^0}{0!} + \frac{E^1}{1!}$$

Working this out for our call-centre number produces:

$$1 + 5 + \frac{5^2}{2!} + \frac{5^3}{3!} + \frac{5^4}{4!} = 1 + 5 + \frac{25}{2} + \frac{125}{6} + \frac{625}{24} = 65.375.$$

Putting our top and bottom layers of algebra together finally gives the formula for Erlang-B, and provides the all important value of P as:

$$P = \frac{\frac{E^n}{n!}}{1 + E + \cdots + \frac{E^n}{n}}$$

For our call centre, where P defines the number of callers not getting through, these figures and Erlang's formula produce a result of:

$$\frac{26.042}{65.375} \approx 0.4$$

That means around 40 per cent of callers will find the line

busy. Remembering that our (no-doubt disappointed) call-centre manager wanted a target of 0.05 (5 per cent) failures. Further analysis suggests that he would need to employ 9 phone operators rather than 4 to bring P below that threshold.

It is this sort of practical insight that makes Erlang's analysis so valuable. Complementing it is a handy rule established by John Little in 1961, relating the number of people in a queue and the length of time each of them has to wait. If the number of people waiting to be served is on average A, and each person waits around W minutes, with around J new people joining the queue each minute, then Little's Law tells us that the average number of people in the queue is given by:

$$A = W \times J$$

In a more modern call centre than the one we've been dealing with, which does have call-waiting technology, A will be precisely the incoming traffic, measured in Erlangs. So Little's Law tells us how to calculate this quantity from the number of people who call up each minute J, and the average waiting time for each caller, W.

Swapping beds and switching packets

Through results such as Erlang-B and Little's Law, the science of queuing has become central to the free-flowing nature of much modern life, in business, industry and computer science. In the context of call centres, technology (notably call-waiting facilities) has overtaken Erlang-B, but it is still very relevant in other contexts, such as predicting blockages in telephone exchanges and similar systems. This includes environments beyond telecommunications – hospital management is a good example. If administrators need to know the likelihood of running out of beds for patients, that will be exactly the value of P provided by Erlang-B.

But the Danish engineer was not finished yet. He also devised a more complex elder brother, Erlang-C, which does incorporate call-waiting. From the same starting points – the volume of incoming traffic and the number of call-centre workers – Erlang-C determines the likelihood that a caller will have to wait for service. Using the same figures as in our call-centre example above, it is virtually certain that callers will have to wait, but with upping the number of staff to nine ($n = 9$), that probability drops to 8 per cent. And Erlang-C has a highly fruitful life beyond call centres too. In the Information Age it is encountered by many of us daily in the form of packet switching, which operates on computer networks and 3G and 4G systems, determining how fast we get access to our webpages. Rather than communicating through continuous streams of binary code, networks often send and receive 'packets' of data, protected by error-correcting codes (see *Avoiding bad language*). These packets can be mathematically modelled as callers to a telephone exchange, and the speed of network will depend on, among other things, the likelihood of queues forming – exactly the number provided by Erlang-C.

From the apparatus of landlines and telephone exchanges to the modern unwired world of mobile phones, smart phones and WiFi, much has changed in the century since Erlang first analysed the humble queue. But in the 21st century his name is alive and well – indeed, more central than ever to the way and speed with which we interact.

AVOIDING BAD LANGUAGE

Correcting mistakes in
digital communication

Arguably, life in the Information Age has become more fragile than day-to-day existence was in simpler times. While the modern connected world of email, video-links, text-message, social networking and the World Wide Web has brought huge benefits, and reduced the world to a 'global village', it brings with it a dependence on the seamless 24/7 functionality of complex computer systems. Increasingly we devolve authority to online technologies to manage our lives. The concomitant fact is that – to take the burgeoning market in internet banking as one example – where once a human or technological error might affect the account holders at a single branch, nowadays a glitch can bring an entire bank's operations into chaos, a scenario with several recent examples. The stakes become even higher when one considers matters of national security or the operations of government and health services. We are both more powerful, and more vulnerable, than ever before.

The human potential to err, will, of course, always be with us. For all their sophistication, the tools of modern life are not perfect: software contains bugs, networks crash and devices malfunction. And of course, not all errors happen by accident: hackers, vandals, thieves and virus writers may also leave their unwanted imprints on our precious code. So an important goal in modern computer science is to find ways to improve our means of transmitting information, making it resistant to errors creeping in. Since, by 'information' we are talking, in computer terms, about streams of binary digits – better known as 'bits' – we are already in the world of mathematics.

A world of zeroes and ones

Bits are the components of the binary – base-two – number system. Every bit takes the form of one of two symbols, 0 and 1, and today's digital information is encoded into long strings of them. If an error creeps in and a single bit gets flipped by accident (say 111001101 … instead of 101001101 …), this could be potentially disastrous: imagine credit card numbers or medical records becoming corrupted. So how can we mitigate against this possibility?

The most naive approach is to repeat each bit – say three times so that 101001101 would become 111000111000000111111000111, so if a single 0 or 1 gets flipped the receiving software can scan it, identify whatever is inconsistent and correct it automatically to the majority value. In simple terms, this illustrates the idea of an error-correcting code. But the exact mechanism leaves a lot to be desired. To begin with, this code is only secure at the level of single, isolated errors, so if two bits in succession get flipped it would be defeated. One could extend the lengths of the repeated blocks from three to five or seven units, but the drawback is that the string starts to become inconveniently long.

Luckily, more elegant, less cumbersome, solutions are possible.

A 'parity bit' offers an alternative approach to guard against errors. At the end of the message, this single new bit is tacked on. Whether it will be a 0 or 1 will depend on a rule: if the total quantity of 1s in the original message is an even number, 0 is added; if the total is an odd number, then the parity bit is 1. For example:

	Original string	Parity bit
Message 1	000101101	0
Message 2	101001101	1

On arrival, the receiving software checks the calculation. This system too is far from perfect. If two bits – or any even number of bits – get flipped, the mistake will not be picked up. And even if an error is detected, the parity bit will not direct us to what exactly has gone wrong. Nevertheless, this system does offer one striking advantage: it only extends the length of the message by a single bit. All the same, something more sophisticated will usually be preferable.

Checking the sums, and byting the nettle

Many computer users are familiar with the word 'checksum', even if a little hazy as to what it means. Essentially, checksums extend the philosophy of the parity bit but their starting point is that binary code, as well being the language of digital information, is also an alternative way to write out whole numbers from our more familiar decimal system. Think of old-fashioned schoolroom addition, where children learned to arrange numbers in columns representing units, tens, hundreds, etc., with the number of each expressed using the symbols 0–9. Binary works in exactly the same way, but the columns represent units, twos, fours, eights, etc., and the number of each can be expressed using only the symbols 0 and 1. A dictionary for translating between decimal and binary would begin:

Decimal	0	1	2	3	4	5	6	7	8	...
Binary	0	1	10	11	100	101	110	111	1000	...

Checksums typically use a block of eight bits, called a 'byte'. Leaving aside its meaning within the data, each byte can be interpreted as a number between decimal 0 (or binary 00000000) and decimal 263 (that is binary 11111111). Splitting a message into bytes, one can add up all the numbers represented by each byte, and append the total (in binary) to the end.

We can follow this by starting with a string 16 bits long – 1010011010010001 – which can therefore be separated into two bytes (10100110 and 10010001), which represent the decimal numbers 166 and 145. They total 311, which in binary is 100110111. To limit the amount of the extra data, we can add on just the final byte of this number, 00110111 – the parity byte. Appending this to the end of the string, the receiving computer can then verify the data by – as the name suggests – checking the sum.

This type of byte checksum is considerably more robust for detecting errors than a single parity bit, as it picks up many more errors. On its own, however, it is not invulnerable. Should some disaster cause the whole string to come through as zeroes, followed by a checksum of 00000000, the receiving computer will accept the integrity of the information with potentially catastrophic consequences, because the checksum is technically accurate. A way to apply a double-check is for the encoding and receiving computers each to flip the checksum's bits, so that – for example – a checksum of 00110111 would instead be written as 11001000. In the case of our all-zeroes calamity, the string will be rejected, because the correct checksum of 0000000000000000 should now be 111111111. This is standard practice, but we can ignore it for the purposes of the rest of our discussion.

Overall, checksum operations do offer the ability to spot the vast majority of coding errors. If an error creeps into the data, and corrupts the bits within a stretch at most eight bits long, the checksum will detect the error. For longer stretches of corrupted data, the checksum still has a probability of around 99.6 per cent of picking up the error, and this can be increased to 99.998 per cent if the checksum is performed using a block of two bytes (16 bits) instead of one.

Hamming codes

Clearly, checksums are very efficient, and variants on the scheme above are commonly used when downloading files from the internet. However, the checksum is not technically an *error-correcting* code – it spots something awry, but cannot fix it. There are other systems that can actually put errors right when they creep in, and the earliest was the Hamming Code, devised in the 1940s by Richard Hamming while he was working for Bell Telephone Laboratories.

This idea deploys several parity bits, in a clever way, and it is worth a little effort exploring one example to convey the principle. Suppose we want to send a message seven bits long, say 1011001. The string we send is going to insert parity bits into positions 1, 2, 4, 8, etc. – in other words, the powers of two. If we name the four parity bits needed for our seven-bit string P_1–P_4, they will therefore occur in the following positions to give us a new string reading: P_1 P_2 1 P_3 0 1 1 P_4 0 0 1.

Each of these parity bits applies just to a particular range of positions, and in our example we now have 11 positions. The correspondences are clearest if we write out the positions in binary:

Position in decimal	1	2	3	4	5	6	7	8	9	10	11
Position in binary	0001	0010	0011	0100	0101	0110	0111	1000	1001	1010	1011
Data bit	P1	P2	1	P3	0	1	1	P4	0	0	1

The first parity bit (P_1) applies to those positions whose binary description ends in a 1, that is to say positions 1, 3, 5, 7, 9 and 11. Ignoring position 1 (which contains P_1 itself) we find that the *data* bits corresponding to these positions read: 10101. This string contains an odd number of 1s, so $P_1 = 1$.

Similarly, P_2 is calculated on those positions whose penultimate bit is 1, meaning positions 2, 3, 6, 7, 10 and 11. These correspond to data bits 11101, with an even number of 1s, making $P_2 = 0$. Continuing in the same way, P_3 depends on positions whose third-from-last bit is 1, which are 4, 5, 6 and 7. The data subsequence here is 011, so $P_3 = 0$. Finally, P_4 depends on positions 8, 9, 10 and 11, which yield 001, meaning that $P_4 = 1$.

When these figures are computed, the string that is actually sent combines our original seven-bit string (1011001) with our four parity bits (1, 0, 0, 1) inserted at the right positions, to make 10100111001.

The Hamming Code may seem a rather elaborate procedure. The point of it is this: each position now lies within the scope of a unique collection of parity bits. For instance, position 7 falls within the remit of P_1, P_2 and P_3, while position 10 is governed by P_2 and P_4. In fact, the parity bits that apply to a given position can read directly from its binary description: in binary, the decimal number 7 is 0111. Because this has 1s in the first, second, and third positions (counting from the right) we can tell that it falls under the sway of P_1, P_2 and P_3. Similarly, decimal 10 is 1010 in binary, and its 1s in the second and fourth positions direct us towards P_2 and P_4.

This makes identifying and correcting errors much easier. If, on the data arriving, P_1, P_2 and P_3 register errors while P_4 is correct, this strongly suggests that the bit in position 7 has gone wrong, and it can then be flipped to put it right. On the other hand if P_2 and P_4 are wrong while P_1 and P_3 are correct, then suspicion falls squarely onto position 10.

With the addition of a final overall parity bit applying to the whole string, Hamming codes have a certain robustness: they can reliably correct single errors and detect (though without correcting) double errors. What is more, for longer strings the number of additional parity bits is very modest compared to the total length of the original string. In a string of 2^m bits, only $m + 1$ of them will be parity bits. For example, if we assume

$m = 8$, a string of length 128 bits need contain only 9 parity bits, leaving 93 per cent of the total space available for 'real' data. This rate improves the longer the string becomes.

Hamming balls and binary codewords

The Hamming Code was the first error-detecting code, and its elegance and efficiency mean that it is still in use today in systems liable to random individual errors. Since Hamming's time, though, the field has grown in size and mathematical sophistication; a variety of error-correcting codes are now used in all manner of technologies, from protecting computers against viruses to allowing CDs and DVDs to function after being scratched.

To analyse the many types of codes in use today, it pays to adopt a more general approach – and this brings an unexpect- edly geometrical flavour to the subject. Starting with two binary 'words', such as 1100 and 1001, the Hamming distance between them is the number of bits that need to be flipped to turn one into another. In this case it is two, the second and last bit needing to be changed.

Using this unusual notion of distance, we may now contem- plate geometrical objects such as spheres. Just as an ordinary ball with radius of 2 centimetres is defined to be the set of all points within 2 centimetres of the centre, so we can talk about the 'Hamming Ball' of radius 2 centred on a binary word such as 1100. This will include other words, such as 1101 and 1001, whose Hamming distance from the centre is no more than 2, but not those such as 0111 which would require 3 or more bits to be flipped to turn one into the other.

Now, whenever words are communicated, errors may creep in, so if we suppose a word we label W is sent across a channel in which up to r errors may occur, we could also express this by saying that the word that arrives with the recipient will lie within the Hamming Ball centred on W with radius r.

To add to the picture, we might now suppose we wish to send a message composed of sixteen binary words, each four bits long: 0000, 0001, 0010, 0011, 0100, 0101, 0110, 0111, 1000, 1001, 1010, 1011, 1100, 1101, 1110 and 1111. Now, to guard against errors we won't send these directly, but will translate each into a longer form, say of seven bits each. There are, in the binary system, $2^7 = 128$ total words of length 7, so not all of them will be codewords for one of our original words. This surplus is useful for it can serve as a buffer zone against errors.

Being sensible, we don't want assign two of our four-bit words to the seven-bit codewords 0000000 and 0000001, since a single error could then cause fatal ambiguity in the message. Rather, we need to choose the 'dictionary' carefully, to maximize the clear blue water between codewords.

If we were to use the Hamming techniques described above to produce our codewords, we would arrive at 0000000, 1101001, 0101010, 1000011, 1001100, 0100101, 1100110, 0001111, 1110000, 0011001, 1011010, 0110011, 0111100, 1010101, 0010110 and 1111111. In each of these, the original data can be read at third, fifth, sixth and seventh positions of each codeword, with the others being parity bits, exactly as discussed earlier. This is now known as a (7,4)-Hamming code, meaning that it encodes words of length 4 as codewords of length 7.

The useful point with this collection is that the distance

between any two codewords picked from the above collection of 16 is always exactly 3. This means that if a word arrives with just a single error in it, there is no ambiguity: there is only one codeword it could be, because there is only one which is exactly one bit-flip away, the second-closest will be two flips away. We can think of this in terms of Hamming Balls: each codeword has a ball around it of radius 1, which provides a safety buffer, since none of these balls overlap. In fact, for our particular code there is even more that we can say: that these same balls cover the whole space, meaning that every possible string of seven bits is within one bit-flip of exactly one codeword. (It will not always be the case that the balls that separate the codewords from each other also cover the entire space, making this an example of what mathematicians call a 'perfect' code.)

It is never pretty to write about bits and bytes: after a little while, strings of 0s and 1s blend into an undifferentiated mass to the human eye. Thankfully, we can look to the algorithms of computer science to do the legwork, including increasingly sophisticated error-correcting codes. Research continues, and the innovative analysis we have glimpsed here has gone even further, finding new and imaginative ways to pack Hamming Balls together in spaces of different dimensions.

In the end it is not surprising that mathematics underpins information transmission and error-detection; but it *is* surprising the form that this mathematics takes. In particular, it is remarkable that guarding against mistakes creeping in when sending an email should depend on the solutions to extremely delicate questions in multi-dimensional geometry.

AUTOMATA AND ARTICULATION

The mathematics of
robot movement

We have the Czech dramatist Karel Čapek to thank for robots – or at least the word, which he introduced to the world in his 1920 play *R.U.R. (Rossum's Universal Robots)*. Here machine-like humanoids, capable of thought, rebel against their lives of drudgery in the service of humans. Čapek's robots drew on the Czech word *robota*, meaning 'servitude', and while science-fiction authors contemplate intelligent robots getting above their station, in today's world we still associate robots with super labour-saving devices of diverse kinds, all of which follow orders unswervingly, unthinkingly – well, 'robotically'.

In the real world, General Motors developed the first industrial robot, named *Unimate*, in 1961. Today, a myriad of devices fit the description 'robot'. While they may work unthinkingly, the same is not true of the engineers and technicians charged with creating and controlling them. Just as young children learn to coordinate their own limbs to achieve the precise movements required by particular tasks, so today's robot-designers attempt to understand and apply those same principles; indeed there is a whole science of kinematics, filled with deep geometrical and algebraic ideas, which is providing ever more advances in our understanding of robotic movement.

Freedom of movement

We may appreciate the principles through some contrasting examples. Imagine a robotic arm, where one end is bolted to a laboratory desk and on the other end there is some sort of device, say a gripper. This robotic hand might be rigidly attached to the arm. In this case, since the arm is entirely jointless, the hand will be able to perform its intended function only if the whole apparatus happens to be positioned correctly.

With adding just one moveable joint, the arm and hand combination become markedly more useful. If we suppose it has a ball-and-socket joint at its base and the rod is 1 metre long, the collection of positions accessible to the hand (known as its workspace), as it moves as far forwards, backwards, left and right as it can go, will form a hemisphere of radius 1 metre. Unfortunately it is only the *surface* of this hemisphere that is accessible. Because the rod is rigid – it has no elbow or wrist – the hand will not be able to draw back or to reach down to a position, say, 0.5 metres from the base. To fill this region, we need a second joint, perhaps half-way along the rod. It might be a hinge joint, like a human elbow, which can bend one way but not the other. If so, then flexing the artificial elbow will cause the hand to sweep out a semi-circular path in space.

As the human body shows, joints come in many types. Nevertheless, there are two fundamental principles at work. The first is rotation, as in the movement of an ordinary door. The axis of rotation in this case is represented by the vertical line through its hinges. Knees and elbows are living examples of such hinge joints in the human body. But a drill bit operates on the same principle – it appears different, however, because its axis runs directly through the bit itself. And thus a human neck or shoulder is also rotational. The second principle is translation, meaning travel in a straight line. A telescopic extension is one example of translation.

Not every movement is covered by these two possibilities. But it is a fundamental geometric fact that every so-called rigid motion can be decomposed into some combination of rotations and translations. For instance, if we keep our arm straight, point a finger ahead, and draw a circle in the air, we are performing two rotations at the same time: around both a vertical axis and a horizontal axis. Similarly, a screwing action with a drill bit is a combination of translation along a straight line, as the drill bit goes deeper, with rotation around the same axis as the drill bit turns.

Clearly, joints and limbs – robotic or human – have varying amounts of room for manoeuvre. This can be made precise by the notion of degrees of freedom. If we cut the hand off our robotic arm and also separate it from the desk to leave the rigid arm floating freely in space, then there are now six fundamentally different ways in which it can move, known as its six degrees of freedom. In terms of translational movement, it can slide along a right–leftward axis, an up–down axis, or a forwards–backwards axis. It can also rotate around each of these axes. Importantly, every other motion, such as translating or rotating around a diagonal axis, can be expressed as a combination of these six.

If we now add a hinge-joint in the middle of the arm, we

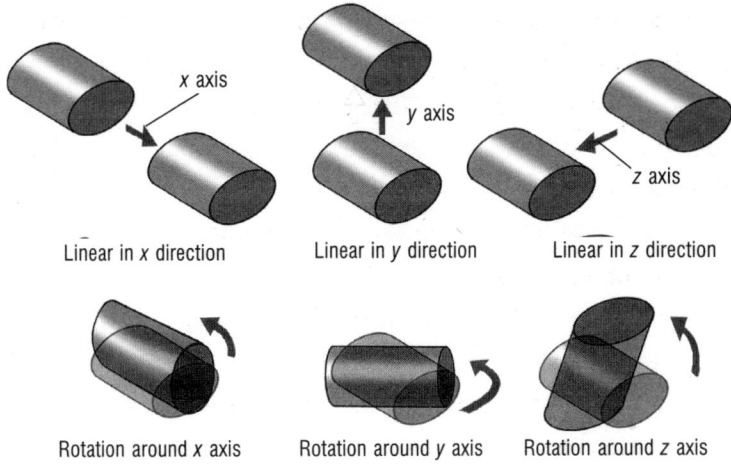

Linear in *x* direction   Linear in *y* direction   Linear in *z* direction

Rotation around *x* axis   Rotation around *y* axis   Rotation around *z* axis

A solid object floating in space has six degrees of freedom.

increase its degrees of freedom to seven, since in addition to the previous six the hinge can now also open and close. But a ball-and-socket joint gives even more flexibility, with three extra degrees of freedom: it can rotate like a drill bit, or close like a hinge in two perpendicular directions. So, a free-floating arm with a ball-and-socket joint in the middle delivers a total of nine degrees of freedom.

Let's consider this more carefully. If our robotic arm has its shoulder fixed to the desk and its hand free, then its total degrees of freedom can be calculated easily: add up those for each of the joints, giving a total of f, say. Alternatively, if both ends are unconstrained allowing the arm to float in space, the total is $f + 6$. These are examples of the *mobility formula*, and in general it says that if a system has n parts, connected by j joints whose individual degrees of freedom add up to f, then the number of degrees of freedom of the entire system is:

$$6n - 6j + f$$

So a robot arm consisting of two rods joined by a hinge joint (one degree of freedom) and attached to the desk with a ball-and-socket joint (three degrees of freedom) has total degrees of freedom of $(6 \times 2) - (6 \times 2) + 4 = 4$. If we place the shoulder joint on a cart travelling along a rail, so it can slide backwards and forwards (one degree of freedom), the degree of freedom increases to $(6 \times 3) - (6 \times 3) + 5 = 5$. If we detach the (rather versatile) cart from the rail and allow it to float freely through space, this goes up to $(6 \times 3) - (6 \times 2) + 4 = 10$.

How many joints?

This sort of thinking matters to anyone designing robotic limbs. Suppose we return to our robotic arm clamped to the desk, perhaps now with a laser pointer for a hand. A basic requirement would be for the hand to access every point within the hemispherical workspace, but beyond this we might also want the laser to be able to point in every direction from every point. This means that the hand must have its full complement of six degrees of freedom, so we need to figure out how many joints are needed to achieve this.

With one end attached to the desk, the number of joints (j) and the number of links (n) will be the same, so $6n - 6j = 0$, and the mobility formula just boils down to the number f: the total of adding up the degrees of freedom of every joint. To achieve our aim, we need this number to be at least six. Typically, industrial robotic arms are built from joints of a simple type, each with just one degree of freedom, meaning hinge joints and rotating elements rather than ball-and-socket types. For this reason, it is standard for today's industrial robots to have exactly six joints.

The human arm, by contrast, has seven degrees of freedom: three in the shoulder, one in the elbow, and three in the wrist. It may appear the robot is therefore more limited, but in fact one might say that the human arm is over-endowed. If we hold our arm out straight in front of us, with our hand horizontally flat, palm downwards, we can roll our shoulder outwards by 90 degrees to leave our hand vertical, as if for a handshake; it is then possible to flatten our hand once again using just the wrist. In this position, opposite rotations in our shoulder (outwards) and wrist (inwards) cancel each other out.

Kinematic conundrums

Of course, not just any six joints will do to create a successful robotic arm. A sequence of six hinge joints along parallel axes would be constrained to single flat plane. So just counting the degrees of freedom is not enough, which is where the more sophisticated science of kinematics enters the story. Should our robotic arm, attached to the table, consist of two rods, there would be four elements involved – the shoulder joint, upper arm, elbow and lower arm. If we specify the position of each joint, and the shape of each link, in what location will the hand end up? This is a question for *forward* kinematics.

The equations that govern its behaviour are of the form

$$M = M_1 \times M_2 \times M_3 \times M_4$$

Here M_1 describes the state of the first item in the chain, the shoulder joint. Then M_2 represents the second, meaning the upper arm, M_3 the elbow, and M_4 the lower arm. Putting these together gives us M, what we want to know, the resulting position of the hand. The more joints the arm has, the longer this equation will inevitably be. We can push the geometrical analysis a step further by observing that each of M_1 ... M_4 is a rigid motion and therefore must be expressible as a combination of a translation (e.g. T_1) and a rotation (e.g. R_1), so, for example, $M_1 = T_1 \times R_1$ and the kinematic equation for a two-joint arm becomes

$$M = T_1 \times R_1 \times T_2 \times R_2 \times T_3 \times R_3 \times T_4 \times R_4$$

This may seem longwinded, but it does illustrate how subtle algebraic considerations become involved in the science of robotics. The version above is valid for *any possible* two-joint arm (and extends in an obvious way to arms with more joints). In any specific arm, however, several of these terms will turn out to be trivial. To see how this works, suppose our arm comprises two rods, each 1 metre long, connected with one hinge joint at the elbow, and attached by another hinge to the desk. If the shoulder at the desk flexes to the right 60° to the horizontal, and the elbow then opens downwards by 90°, where will the hand be?

The answer is ultimately given by the kinematic equation, but a more intuitive way of thinking about it is to imagine an ant walking along the arm, translating and rotating at each juncture. It starts at the shoulder, whose coordinates we'll take as (0, 0), facing horizontally rightwards. Since the shoulder does not move laterally, T_1 has no effect. But R_1 spins the ant upwards to an angle of 60°. Next, T_2 tells the ant to walk forwards 1 metre along the upper arm. Using trigonometry – the science of angles and distances – the coordinates of the point our ant reaches are

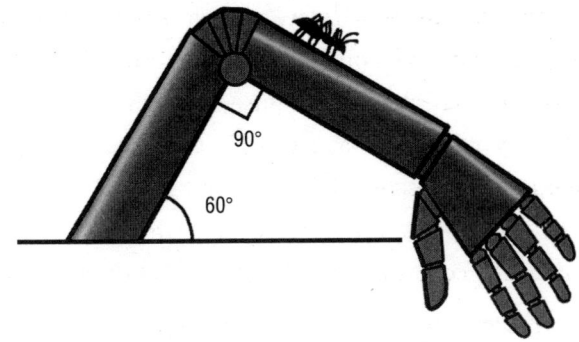

The kinematic equation for a robotic arm can be thought of as a set of instructions for an ant crawling from the base to the hand.

$$\left(\frac{1}{2}, \frac{\sqrt{3}}{2}\right)$$

The ant's journey along the upper arm is completed by the rotation R_2, but because the rod is perfectly straight, this does nothing to our ant. Since the elbow does not move laterally, the same goes for T_3. At R_3, however, the elbow rotates downwards by 90°, leaving our ant looking downwards at an angle of 30°. The final step (since R_4 does nothing) is T_4, where the ant walks forwards another metre. This takes it to the hand, where our ant is pointing downwards at an angle of 30° and located approximately at the point (1.4, 0.4), or, more precisely, at the coordinates

$$\left(\frac{1+\sqrt{3}}{2}, \frac{\sqrt{3}-1}{2}\right)$$

The ant's final state tells us exactly the location and angle of the hand.

Forward kinematics, therefore, answers questions about the position of the hand when the links are of given proportions and the joints are in certain positions. But often the question is the reverse: if we want the robotic hand to be in a certain location, facing in a certain direction, what configurations do the various joints need to adopt? This is a matter of *inverse* kinematics, and it is the fundamental problem that a robot's controller needs to solve when instructing it on how to perform useful tasks.

While forwards kinematics always produces a unique answer, in inverse kinematics there may be multiple ways for the hand to reach its goal. For human arms, we had a taste of this in the redundancy provided by the seventh degree of freedom. Furthermore, the number of ways we might have our hand flat at arm's length in front of us is, perhaps surprisingly, infinite, because we can rotate our shoulder outwards by any angle x (up to some limit), and then rotate our wrist inwards by the same amount. Since the number of possible values of x is infinite, this particular kinematic equation has infinitely many solutions.

Things are not quite so open-ended for robotic arms with the standard six degrees of freedom, and there will typically be a finite number of different ways of setting the joints to give the desired result. In fact, for a standard six-joint arm this number will be no more than 16, a fact first proved by Eric Primrose in 1986.

Flight simulators and spaceship docking

Thus far our robots have been of the *serial* type – a sequence of links and joints, one following the other. This is standard for the sorts of robotic arms used in the construction industry, but in other contexts robots of different designs are needed, and one of the best known (and most useful) is the so-called Stewart platform, also known as a 'hexapod'. This six-legged robot, supporting a table-top platform, was originally designed in the 1950s by Eric Gough, before the kinematics of the robot were laid out in a paper of 1965 by D. Stewart.

To assess its degrees of freedom, we need to look at its vital statistics. Each leg comes in two pieces, which, along with the platform, give the machine a total of $n = 13$ moving parts. Each leg has a joint at the base, another in the centre, and one at the top, making a total of $j = 3 \times 6 = 18$ joints in the entire system.

Each of these joints has two degrees of freedom (all of which are purely rotational, except for the ones in the centre of the legs which are telescopic). So adding up all these individual degrees of freedom gives $f = 2 \times 18 = 36$. Applying the mobility formula from above to calculate the robot's total degrees of freedom produces the equation:

$$6n - 6j + f = 6 \times 13 - 6 \times 18 + 36 = 6$$

This magic value suggests the Stewart platform's real usefulness: an object sitting on the platform can be manipulated in all six degrees of freedom. It can be slid forwards and backwards, leftwards and rightwards, and lifted up and down; it can also be tilted to the front, back, right and left, as well as swivelling clockwise or anticlockwise. Despite this flexibility, the Stewart platform also has the advantage of great rigidity: the platform is wobble-free. What is more, the inverse kinematics of this robot works comparatively smoothly: to manoeuvre the platform into a given position all that is needed is to adjust the extensions of each of its six legs – the remaining rotational components simply fall into place.

Their flexibility and strength mean that Stewart platforms have been used in numerous contexts: Stewart conceived the design as a flight simulator, and large Stewart platforms continue to be used this way for pilot training. More people will be familiar with them from riding in virtual roller-coaster pods in theme parks and the like, where customers enter a small cabin atop the platform, take their seats, and watch the journey (to the stars, down a mine, into the air) on a screen as they are tossed this way and that by the robot. More seriously, Stewart platforms play major roles as precision tools in industry – from crane hoists to bridge building – and space technology, where, among a varied repertoire of tasks, they have contributed to the docking mechanisms at the International Space Station.

In creating the robotic devices of today and tomorrow, and despite the fairly straightforward introduction given here, sophisticated techniques of numerical analysis (see *Rain or shine?*) may be needed to solve a robot's particular kinematic equation, and to unearth the correct orientations for all its joints. Indeed, it is no exaggeration to say that one of the major challenges of robot design is to build robots whose kinematic equations can be solved quickly and easily.

The robotic implications of mathematics go back at least to the 18th century, when Leonhard Euler showed that rigid motion was a combination of translational and rotational movement. It is clear that mathematicians and robots will need to keep close company for some time yet.

HOT STUFF

The mathematics of
energy and entropy

There are few more changeable phenomena than energy. It surrounds us and powers us, though we may only take particular notice when there is too much of it, or not enough: a rampaging toddler tearing around the house, or the central heating not working properly. Its variety of forms include the heat of a fire, the light and sound of an electrical storm, the kinetic energy of a car, the gravitational potential of an object suspended high above the ground, or the chemical energy in a tank of petrol. And management of energy, from our own body's exercise regime (or lack of one) to a nation's resources, is central to modern societies.

The science of energy is thermodynamics, and to understand it a few mathematical principles are needed. Of particular interest is the concept of 'entropy', which has gradually expanded its remit to become one of the most important notions in science, helping theoretical physicists contemplate some of the strangest phenomena in the universe. More surprisingly, it also connects thermodynamics with an ostensibly unrelated discipline – the information theory that underpins the internet.

Dynamics and disorder

Scientists have discovered many ways by which energy is converted from one form to another. To take one example, the internal combustion engine – possibly the defining technology of the early 20th century – relies on the fact that energy stored in the chemical bonds of petroleum molecules will convert to heat and kinetic energy when the fuel is burned. The First Law of Thermodynamics underlies the process, asserting that energy may be transferred from one form to another, but it can never be created or destroyed. That is to say, the total energy in the universe will always remain constant. This also goes by the name of the *law of conservation of energy*. The same is true for any closed system, meaning one which is self-contained and does not lose or gain energy from the outside. This does not include our own planet, of course, which is far from a closed system, since it receives almost all its energy from the sun. The Second Law of Thermodynamics is somewhat subtler, and can be expressed in various ways, such as the statement that, whenever energy is used to produce physical movement, the process can never be 100 per cent efficient. Some energy will always dissipate as heat, sound, light, etc.

Attempts to defy these laws have a long history. The laws outlaw perpetual motion machines, yet that has proved no disincentive to ambitious inventors over the centuries. One early prototype tried to use the energy from a waterwheel to pump water up to a reservoir, which then drained under gravity to power the wheel. A little inspection reveals the flaws in this (and indeed in all) perpetual motion machines. Since, by the First Law, the device cannot create any new energy, it means that the machine cannot power anything beyond itself. However, the Second Law guarantees that it will gradually run down even so, because some energy will be lost in the process. The two laws are sometimes facetiously paraphrased as 'You can't win' (First Law) and 'You can't break even' (Second Law).

An alternative early expression of the Second Law was Rudolf Clausius's assertion in 1856 that there can be no physical process whose only net result is to transfer heat from a cool object to a warmer one. A somewhat more profound variant, also originally framed by Clausius, is the statement that within any closed system the entropy of the entire system will always increase over time. 'Entropy' is commonly described as a measure of *disorder* of the system; but caution is needed here, since disorder may be in the eye of the beholder: a piece of modern art may appear random or disordered but is in fact a carefully honed product of long deliberation; a teenager's seemingly messy bedroom might obscure the fact that its occupant is actually in full control the whereabouts of all its contents, so there is some method in the madness. The notion of entropy has developed over time, becoming one of the most important in science. But it is clearly unacceptable for the science of thermodynamics to depend on subjective interpretations of 'order' and 'disorder'.

'Entropy' as a description can only be used at the level of the *macrostate* of an object. For a canister of gas, its *macrostate* is a description of large-scale properties such as pressure, temperature, volume, and so on. Conversely (leaving aside quantum effects), an attempt to describe the *microstate* of the gas would entail a detailed account of the motion and position of every molecule within it. Entropy is not meaningful at this level.

The entropy of a macrostate increases with the number of possible microstates that could produce it. For example, a vat of gas in which the fast-moving, hot molecules are all on the left, and the slow, cool molecules are on the right, has a much lower entropy than one where the molecules are mixed together. This is not in the eye of the beholder: it is a mathematical observation, because the number of ways in which the latter can happen far outnumbers those of the former.

To illustrate the point, imagine our hot and cold particles as 5 blue balls and 5 red ones arranged in a single row. Now, the total number of ways of arranging the balls, using the factorial function (as for counting queues: see *Hold the line, please!*), is $10! = 362,800$. Out of this total, only a relatively few yield the special ordered configuration where all the red balls are on the left and the blue balls on the right. In fact the number of arrangements that fit that description is $5! \times 5! = 14,400$, or around 4 per cent of the total number of possibilities. If the balls are the other way round (all blue on the left and red on the right), this accounts for another 4 per cent of the total. So these macrostates have low

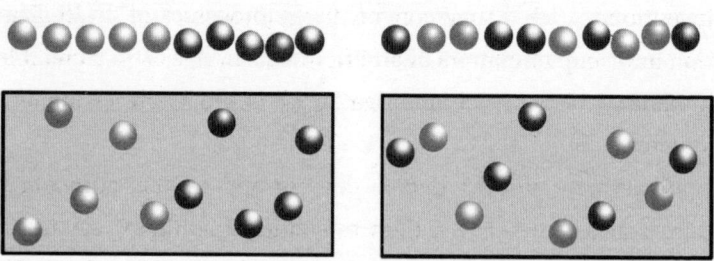

Lower entropy configurations are on the left, while the more disordered higher entropy configurations are on the right.

entropy compared with the much larger number of cases (92 per cent) in which the balls are to some extent mixed up.

Enumerating entropy

Entropy is traditionally denoted by the letter S (the letter E having already been bagged by energy). For a given macrostate, the value of S depends on the number of compatible microstates, which we can call W. Of course, we could just define the entropy to be equal to this number: $S = W$. But it turns out to be more convenient to take its *logarithm* and define $S = \log W$. So if a canister of gas is compatible with 8 (= 2^3) different microstates, then its entropy is 3, while another where $W = 128$ (= 2^7) has an entropy of 7. (For more on logarithms, see *The rise of* homo economicus.) There is a technical assumption here that all microstates are equally probable; it is possible to dispense with this assumption, which leads to a slightly more complicated formula, but we will put this issue to one side.

The introduction of the logarithm may seem an unnecessary complication, but it has some convenient consequences. The essence of logarithms – indeed the very reason for their invention – is that they convert tricky processes of multiplication into simpler matters of addition. This brings us to a credible method of *adding* entropies together when we combine different systems.

If we were just to define the entropy as the number of microstates (i.e. $S = W$), then this kind of addition would not work. Imagine two canisters next to each other, the first with 8 compatible microstates, and the second with 128. If we fix a microstate for the first canister, the second still has the full gamut 128. So, to calculate the number of microstates for the combined system, we need to multiply the two values: $8 \times 128 =$ 1,024. Taking the logarithm instead, and defining $S = \log W$ allows us to get around this problem. Now, the two canisters have entropies of $\log 8 = 3$ and $\log 128 = 7$, while the combination has an entropy of $S = \log 1,024 = 10$. It is no coincidence that this is the total of the entropies of the two individual canisters: $3 + 7 = 10$. (Below we will see another way to account for the presence of the logarithm, in the unexpected relationship between energy and information.)

Thus far, then, we have the entropy of a macrostate as expressed by $S = \log W$, where W is the number of compatible microstates. But there are two further technical details which mean that a physicist's presentation of entropy appears superficially different from the mathematician's. The first is that, for historical reasons, physicists traditionally use the *natural logarithm*, written $\ln W$, which means the logarithm whose base is the mathematical constant known as e (around 2.718), rather than the base-two we are mostly using in this book. This is little more than a cosmetic change.

The second factor is that in any realistic scenario the number of microstates W is likely to be *gigantic*. So, to end up with a manageably sized measure of entropy, it is useful to scale down the whole thing. The fixed number used to do this is known as Boltzmann's constant or simply k. Seen altogether, the definition of entropy that a physicist would recognize is given by Boltzmann's formula $S = k \times \ln W$.

Boltzmann's constant k is a tiny number, at around 1.380×10^{-23} joules per kelvin. It can be thought of as a way to relate the energy of a single particle (an element of a microstate) to the temperature of the gas as a whole (a consequence of the macrostate).

The demon in the detail

The Second Law of Thermodynamics states that the entropy of any closed system can only increase over time. But what does this statement mean in terms of microstates? We can return to our two canisters of gas, but now imagine them as two chambers of a single vat. Suppose we heat the two to different temperatures, and then remove the separating wall. At that moment, the hot gas (meaning fast-moving molecules) is on the left, and the cool gas (slow-moving molecules) on the right. It is obvious what will happen next: left to their own devices, the two gases will mix, and after a little while we will have a vat of mid-temperature gas, in which both slow- and fast-moving molecules are evenly distributed throughout. As we saw above, this mixed state has a higher entropy. Ultimately, the Second Law is also a statistical rule guaranteeing that this type of mixing must happen.

But in 1867, James Clerk Maxwell conceived an idea for defying the Second Law, not through a perpetual motion machine, nor indeed a real device of any kind, but through a crafty thought experiment. It began with the vat of mixed gas, with the partition wall in place dividing it into two sections. The twist was to insert a tiny closable window in the wall. Some entity – dubbed Maxwell's Demon – controls the opening and shutting of the window at certain critical moments, to let individual molecules pass from the right-hand vat into the left-hand one, or vice versa. Specifically, the demon will open the window to allow a fast-moving molecule to travel right–left, or a slow molecule to go the other way. But it will shut the window to any fast-moving molecules travelling rightwards, or slow ones moving leftwards. After some time, the left part of the vat will contain hot gas, meaning fast-moving molecules, while the slower ones are on the colder right-hand side. By cunningly opening and closing the little window, the demon has seemingly violated the Second Law of Thermodynamics.

Maxwell's Demon has troubled scientists for many years. The first counter-argument is that the vat on its own cannot be considered a closed system. Today's theorists believe that the key to the paradox is in understanding the system's other component: the Demon itself. Whatever it is, it at least has the means to monitor the speed of molecules, and to open and close the little window. In performing these activities, the Demon must expend energy. Therefore, in general, we can rest easily, confident that the entire system, with the Demon included, will gain entropy at a greater rate than the vat alone loses it.

There is, intriguingly enough, a borderline case. In 1982, Charles Bennett argued that the Demon may *really* be able to perform its devilish duties, without thermodynamic cost, so long as it has an additional component: a memory where it can store all the data it receives about the molecules in the vat. However, any such memory will necessarily have finite capacity, so the Demon cannot continue its task indefinitely. At some point it will have to clear its memory, at which point the entropy of the whole system jumps up in line with the second law of thermodynamics. This brings us to one of the most unexpected – and wonderful – examples of cross-pollination in science.

Entropic extensions

'Entropy' as a word and a concept is not limited to the science of thermodynamics. It also plays an increasingly prominent role in information theory, the science of how data can be communicated efficiently – something that on the face of it has little to do with energy. In the Information Age, this has become a hugely important subject, with error-correcting codes (see *Avoiding bad language*) a prime example of the theory at work. The thermodynamic–informational link comes from conceiving the entropy of a macrostate as a measure of the amount of *information* needed to specify a single microstate. It also offers an alternative explanation for the logarithm we encountered above.

To appreciate this, we might imagine a canister of gas in a macrostate for which there are 8 possible microstates, which we can number 0 up to 7. In fact, it is more convenient to express these numbers in the binary, as opposed to decimal, system, in which case they would be numbered 000, 001, 010, 011, 100, 101, 110 and 111. The crucial observation here is that all these binary numbers are three bits (binary digits) long. So the number of bits needed to distinguish any one microstate from the others is 3. It is no coincidence that this is the logarithm of 8. (Here we are reverting to the computer scientist's binary logarithm, rather than the physicist's preferred natural logarithm.)

In information theory, entropy can be thought of either as a measure of ignorance or of information. In the 8-microstate example above, the information needed to specify a microstate comes to 3 bits. In the 128-microstate example, 7 bits would be needed. In both cases, this number can be interpreted as quantifying our ignorance of the situation, or equivalently, the amount

353

of information encoded within it.

From the information perspective, what does the Second Law of Thermodynamics say? It asserts that no closed system will evolve from one state into another where you can pin down its microstate with greater accuracy. This unexpected convergence of two such different branches of thought is one of the most exciting developments in recent science. It is now a fundamental tool that scientists use to probe the universe, including some of its strangest parts. For there is one possible way that this expression of the Second Law of Thermodynamics might be broken: if information is destroyed. And where in the universe is that most likely to happen? The obvious answer is: a black hole.

We can conduct our own thought experiment, by tossing our high-entropy vat of gas into a black hole. Should the whole system be crushed to nothing, and all its information irreversibly lost, then the entropy of the entire universe has just decreased, violating the Second Law of Thermodynamics. To avoid this conclusion, black-hole thermodynamics has gripped the imagination of physicists in recent years and, luckily for the Second Law, it is now believed that black holes do indeed have entropy, meaning that information absorbed by a black hole is in some sense preserved. Understanding *how* this happens is a major endeavour. The Bekenstein–Hawking formula, named after Jakob Bekenstein and Stephen Hawking, gives the entropy of a black hole as:

$$S = \frac{A}{4l^2}$$

where A is the surface area of the black hole's event horizon – the sphere around it representing the point of no return – l is the measurement known as the 'Planck length' (around 2×10^{-35} metres). Because l is so small, and A is typically very large, the value of S comes out as a very large number indeed.

This hybrid science of thermodynamics and information theory is one of the hottest topics in science today, providing fresh perspectives on all manner of subjects. But no conclusion is more striking or counterintuitive than that implied by the Bekenstein–Hawking formula: that black holes are actually hugely rich bodies of information.

THE PERILS OF PARADOX

Type theory and programming

In the early years of the 20th century, the philosopher Bertrand Russell and G.G. Berry, a librarian working at Oxford's Bodleian Library set the cat among the pigeons with a short phrase: 'the smallest number which cannot be defined in twelve words of English'. Asked to find this number, one might set off on a hunt for a very large quantity indeed, which requires more than 12 words to pin down. The truth, however, is that no such number can exist. The key here is that the phrase itself contains 12 words, yet describes a number that *cannot* be defined in 12 words – so the description automatically fails to apply to every promising candidate!

This teaser, known as Berry's Paradox, is an example of self-reference, a phenomenon that thinkers in the fields of logic, linguistics and philosophy have been grappling with for the last century. But there is also a practical side to these investigations. In developing a mathematical theory, or a computer programming language, there must be vigilance to ensure that paradoxes do not creep in. The ways in which people have done this have resulted in some astonishing – and supremely useful – repercussions for humanity's search for mathematical truth.

Looking inwards, looking outwards

Berry's Paradox is an example of what, in another field, the anthropologist Gregory Bateson called a double-bind, 'a confusion of message and meta-message'. For Bateson, such phenomena were important in psychology, including to the understanding of schizophrenia, though in practice that idea has proved hard to test.

Logicians know this phenomenon as 'impredicativity'. We can contrast it with its opposite – a non-paradoxical predicative description such as 'the smallest number not expressible using eleven symbols from the following alphabet: 1, 2, 3, 4, 5, 6, 7, 8, 9, $+, -, \times, \div$ and $\wedge$'. (The symbol '$\wedge$' here stands for taking powers, so $2\wedge3$ is the same as 2^3). To work this one out, we could begin by considering all possible arrangements of the eleven symbols and find the biggest number possible, which would be $9\wedge9\wedge9\wedge9\wedge9$, so the smallest number *not* expressible using eleven symbols is $9\wedge9\wedge9\wedge9\wedge9+1$.

Of course, a similar approach cannot work for Berry's Paradox. But in a lighter vein, the physicist John Baez came up with a predicative variant, which we might call Baez's Non-Paradox: 'the smallest whole number that cannot be described in fifteen words without use of self-reference' – using, of course, 15 words. In principle (though highly challenging in practice), an exhaustive search might eventually reveal a numerical value here.

Berry's conundrum notwithstanding, self-reference, or impredicativity, does not have to be inherently paradoxical. Indeed, Frank Ramsey pointed out that many commonplace constructions such as 'the tallest person in the room' are impredicative, at least to a certain extent. If we call this person *T*, then the definition of T relies on an analysis of all the people in the room, a collection that includes T himself. The mathematics here is simple – just to pick the maximum from a collection of heights. But on what basis can we tell the difference between this sort of harmless impredicativity and the impossible double-binds of the type exhibited by Berry? In the early 20th century, a fierce debate about such questions arose among those wanting to provide mathematics with a solid grounding.

Type casting

Seeking a way out of this maze, Bertrand Russell and Alfred North Whitehead developed the new approach of type theory. Originally it was a purely mathematical idea, but decades later it would come to have a greater importance, in the design and understanding of computer programming languages.

The basic premise is familiar from fiction. Imagine a futuristic virtual-reality game, but one plugged directly into the player's nervous system. It might be that within the virtual world are further VR games which one can plug into and play – worlds within worlds, reminiscent of the play-within-a-play device employed by Shakespeare, or the nested dreams in the 2010 film *Inception*.

In our example, we can now assign *types* to these games and sub-games, designating each a number. Reality has type 0, the original game has type 1, and sub-games of type 1 have type 2, etc. Similarly, mathematical type theory provides a strict order in which one can define things: first one defines objects of the base type, perhaps numbers. Next, objects of a higher type are defined, which might include sets of numbers; beyond that, sets of sets of numbers, and so on. A fundamental principle is that an object can refer to other objects of a lower type than itself, but not those of equal or higher type, just as a player in the VR world can be affected by events at her own level or lower (if someone unplugs her machine for instance), but is immune to occurrences at higher types.

In their mammoth three-volume work *Principia Mathematica*, Russell and Whitehead's original type theory went some way towards eliminating paradoxes from mathematical logic. But it did so at a considerable cost, introducing an enormous pile of seemingly unnecessary technical considerations. To illustrate why, let's return to contemplating people in the room. Each individual, such as Nicholas, is an object of type 0. 'The collection of people in the room' is then an entity of type 1, and the 'the tallest person in the room' is one of type 2. So the seemingly straightforward assertion that 'Nicholas is the tallest person in the room' comes to adopt a rather complicated type-straddling structure. The victory for logic is that, interpreted in this way, the sentence is no longer impredicative; though whether the cost is worth it is a matter of opinion.

For all its early clumsiness, type theory would later become commonplace, and even acquire a certain elegance of its own. It was a development accompanying the dawn of computer programming – or, more specifically, the invention in 1956 of Fortran, the first high-level programming language, created by John Backus and his team at IBM. This breakthrough allowed a human computer-operator to write instructions in a much more comprehensible and intuitive way, without having to worry about the jumble of binary 0s and 1s churning away in the computer's memory. The descendants of Fortran are too numerous to list, but include languages like C, Java, Python, PHP and Ruby. In these highly conceptual frameworks, the theory of types has come to play a central role.

In almost all programming languages, the expressions 2 and '2' are fundamentally different entities. The first is a number, and the second is an object of a different type: a string. This difference is similar to the difference between objects and the English words that describe them: a guitar is different from the word 'guitar'. Yet in both cases there is an obvious connection between the two, and it is this relationship that makes it possible to retrieve the number from the string, or vice versa.

For today's programmers, therefore, types are not defined just by levels (as in our virtual reality games) but by different *kinds*, of which the above-mentioned strings (S) and numbers (N) are two such. From these we can construct a new type, written $S \times N$. The objects of this type are pairs [s, n], where the first entry s has type S (that is, it is a string), while the second n has type N, a number. Plugging in some actual values, an example of this compound type is ['3', 2].

An alternative compound type may be written $S + N$ comprising everything that is of type S along with everything of type N – in other words, the full collection of numbers and strings. Beyond this, we can also create a higher-level type of a more complex kind, denoted by $N \rightarrow S$. Objects of this type are *rules* for converting inputs of one type into outputs of another, so in our example of strings and numbers it could be the rule turning 1 into '1', 2 into '2', and so on.

A correspondence course

The constructions described above – and others besides – can be iterated as many times as required, in a variety of combinations, thereby building types of considerable complexity. This is exactly how more complex data-structures are built: arrays, lists, dictionaries, spreadsheets, databases and the like may all be defined via a careful description of the types involved.

Once this method is appreciated, it becomes clear how type theory is an efficient organizing scheme for the many kinds of objects needed by today's computer programmers. But there's more to it than that; in particular, it provides the setting for programming's beautiful relationship with mathematics. While both subjects are rigorously logical, their goals are essentially different: mathematicians deal in proofs, watertight arguments in support of some assertion; programmers design means to perform a particular task. Yet, from 1958 two Americans, the mathematician Haskell Curry and the logician William Howard, realized that proofs and programs were ultimately the same thing, though expressed in very different languages. Their discovery, known as the Curry–Howard Correspondence, was a way to translate the language of data-types into mathematical logic.

The starting point was supremely unexpected: data-types will stand for logical assertions. What is more, the way complicated types are assembled corresponds to the manner in which such assertions may be combined. According to this philosophy, some type A will be interpreted as saying that some assertion a is true. The same is true for another type B with regard to statement b. Things become interesting when we reach combined types: $A \times B$ will stand for 'a and b', while $A + B$ becomes 'a or b'. Other rules describe how the manipulation of types corresponds to statements about logic. Importantly, the type $A \to B$ turns into logical implication: that is, the assertion that 'a implies b'.

Now we might well ask what is to be gained from this strange and seemingly unnatural interpretation. There is a cunning answer: it all hinges on the notion of empty types – a type for which there is no corresponding object. The key rule is that empty types will correspond to *false* assertions, while populated types (with something in them) represent *true* assertions.

If the type $A \to B$ is populated, that means there is some rule for turning objects of type A into objects of type B. If the type A is also populated, that means we have some object of type A to which we can apply the rule, thus giving us an object of type B. It follows therefore that B must also be populated. This thinking gels perfectly with the logical perspective that if 'a' and 'a implies b' are both true, then 'b' must be true too.

This is the essence of the link between the logic of types and mathematical assertions. Populated/empty types represent assertions that are true/false. But the connection runs deeper than this; after all, the obvious way to demonstrate that a type is populated is to exhibit an object of that type. So any object of the specified type will serve as a proof of the corresponding assertion. This is how computer programs, which provide instructions for constructing an object of a given type, can simultaneously function as mathematical proofs.

The satisfying, unanticipated Curry–Howard Correspondence has brought a much deeper understanding of the type theory behind today's programming languages. It has also yielded profound consequences for practising mathematicians, notably in the shape of proof-checking software such as Coq, developed by a team of programmers in Paris led by Hugo Herbelin. Capable of verifying lengthy and complex arguments, Coq achieved its greatest triumph to date with a verification of the celebrated four-colour theorem. This geometrical conjecture states that any 'map' – in the sense of a plane divided into various shapes ('countries') – can be painted with four colours, without bordering countries having the same colour. Although tantalizingly simple to state, the conjecture defied all attempts at proof or counter-example for the best part of a century. That state of affairs ended in 1976, when Kenneth Appel and Wolfgang Haken at the University of Illinois finally provided the long sought-for proof.

There was a catch, however. Their argument was gigantic, containing over 10,000 diagrams and requiring over 1,000 hours of computer time. So how to verify it? Of course, many mathematicians did delve in and were convinced by what they found; but certainty had to wait until 2004, when Georges Gonthier (at Microsoft Research) and Benjamin Werner (at the French National Institute for Research in Computer Science and Control) were able to put Coq to work.

Coq seems an apt note to strike at the conclusion of this book's journey through a few of the many ways in which mathematics has underpinned and enriched human thought and endeavour. Mathematics is the most generous of disciplines, giving of itself unstintingly, to the school-timetabler and the landscape artist, to the medical researcher and the astrophysicist. It seems only fair that among this panoply of applications, mathematics should also generate a few good tools for itself.

INDEX

Quercus Editions Ltd
55 Baker Street
7th floor, South Block
London
W1U 8EW

First published in 2013

A catalogue record of this book is available from the British Library

UK and associated territories: ISBN 978 1 78087 160 8

Illustrations by Patrick Nugent

Printed and bound in China

10 9 8 7 6 5 4 3 2 1